Computational Acoustics in Architecture

International Series on Advances in Architecture

Objectives

The field of architecture has experienced considerable advances in the last few years, many of them connected with new methods and processes, the development of faster and better computer systems and a new interest in our architectural heritage. It is to bring such advances to the attention of the international community that this book series has been established. The object of the series is to publish state-of-the-art information on architectural topics with particular reference to advances in new fields, such as virtual architecture, intelligent systems, novel structural forms, material technology and applications, restoration techniques, movable and lightweight structures, high rise buildings, architectural acoustics, leisure structures, intelligent buildings and other original developments. The Advances in Architecture series consists of a few volumes per year, each under the editorship - by invitation only - of an outstanding architect or researcher. This commitment is backed by an illustrious Editorial Board. Volumes in the Series cover areas of current interest or active research and include contributions by leaders in the field.

Associate Editors

M. Majowiecki
Department of Risorgimento 2
University of Bologna
40100 Bologna
Italy

M. Zador
Faculty of Architecture
Technical University of Budapest
1111 Budapest
Muegyetem rakpart 3.K.II.60
Hungary

S. Sánchez-Beitia
ETS Architectura
Plaza de Onate 2
20009 San Sebastian
Spain

R. Zarnic
Faculty of Civil Engineering
University of Ljubljana
P O Box 579
SI-1000 Ljubljana
Jamova 2
Slovenia

J.J. Sendra
Dept. de Construcciones
Arquitectónicas
Universidad de Sevilla
Avda. Reina Mercedes 2
41012 Sevilla
Spain

Computational Acoustics in Architecture

Edited by

J.J. Sendra
University of Sevilla, Spain

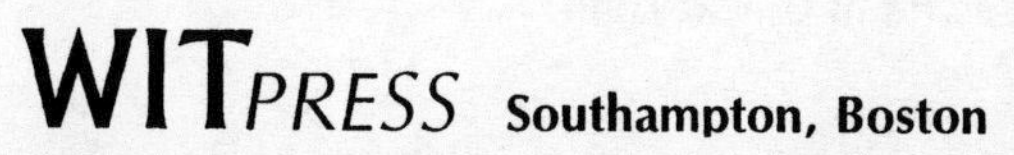

Southampton, Boston

J.J. Sendra
University of Sevilla, Spain.

Published by

WIT Press
Ashurst Lodge, Ashurst, Southampton, SO40 7AA, UK
Tel: 44 (0) 238 029 3223; Fax: 44 (0) 238 029 2853
E-Mail: witpress@witpress.com
http://www.witpress.com

For USA, Canada and Mexico

Computational Mechanics Inc
25 Bridge Street, Billerica, MA 01821, USA
Tel: 978 667 5841; Fax: 978 667 7582
E-Mail: cmina@ix.netcom.com
Mirror site: http://www.compmech.com

British Library Cataloguing-in-Publication Data

A Catalogue record for this book is available
from the British Library

ISBN: 1 85312 557 1
ISSN: 1368-1435

Library of Congress Catalog Card Number: 99-63756

The text of the papers in this volume were set individually by the authors
or under their supervision.

No responsibility is assumed by the Publisher, the Editors and Authors for any injury and/or damage to persons or property as a matter of products liability, negligence or otherwise, or from any use or operation of any methods, products, instructions or ideas contained in the material herein.

Printed in Great Britain by Print in Black, Bath.

Contents

Preface

The scientific discipline known as Architectural Acoustics is but a century old. Its origins are to be found in the studies and experiences of Wallace C. Sabine, which led to the acoustic conditioning of the Boston Music Hall (now the Boston Symphony Hall). With a capacity for 2600 people, it opened to the acclaim of critics and public alike in 1900, the first hall to be built following explicit scientific criteria. Sabine's work, which might be considered the starting point for this new applied science, was reflected in a series of papers, although they appeared in publications in the fields of industry and building, not in academic circles. Previously (1877-78), Lord Rayleigh had published "Theory of Sound" in England, a two-volume treatise which laid the foundations for the theory of Acoustics, though his comments on Room Acoustics did not go beyond generalizations.

Until that time, achieving good acoustic conditions in a room meant relying on guesswork, as well as the experience one gains through trial and error. Sabine's formula for obtaining reverberation time was to Architectural Acoustics what Ohm's law might once have been to electricity, or Hooke's law to elasticity. Any shortcomings to this formula cannot hide one fundamental fact: For the first time, it was possible to carry out a preliminary study of the premises and make a suitable choice and layout of surface materials based on genuinely scientific criteria. Moreover, he succeeded in doing what all scientific research should attempt: Posing new problems.

Following those early studies of Sabine's, considerable work and research has been done with the common goal of learning more about the sound field in enclosures, thus providing a solid scientific base from which to approach this difficult subject that, as M. Schroeder states in the introduction to his chapter, "...lives at the intersection of physical science, engineering and art".

Yet most scientific advances in these last hundred years are due to great breakthroughs in technology. Indeed, highly sophisticated equipment and electronic instruments have become available with which to take acoustic measurements in situ in different rooms, as well as to perform experiments that have helped towards a greater understanding of the workings of the human ear; Sabine's only aid in his research was a stop-watch and his own hearing. Not only has this made precise assessment of a room's acoustics possible, relating objective variables, obtained directly from such measurements, to subjective parameters to do with hearing (speech intelligibility, musical clarity, reverberation, listener envelopment, acoustical intimacy, etc.), but it has also allowed for contrasting the validity of certain theoretical models that have been in use or the suitability of proposals that are currently arising as alternatives.

Due to major advances in electro-acoustics, today we can produce highly sophisticated support systems for the sound levels inside a room, and even create an artificial reverberation or an assisted resonance. The former dates from the 1930s, following the development of radio-broadcasting that led to breakthroughs in the technology of microphones, loudspeakers and amplifiers, mainly for cinemas and conference halls. The latter dates from the 1960s and has proved to be highly effective for correcting acoustic defects in a room and for adjusting acoustic conditions in multipurpose halls, where reverberation times obtained on different frequency bands may be adjusted to those considered optimum according to the uses they are to be given; even to the point of simulating the conditions of an indoor hall for open-air theatres and pavilions.

Among the first notable systems of artificial reverberation used for enclosures was the one adopted by P.H. Parkin in 1964 for the Royal Festival Hall in London, with a capacity for 3000 people, with the aim of increasing its reverberation time (up to 50% in the low and medium frequency ranges), thus correcting what was considered its most serious acoustic flaw. Raising the ceiling (which would have increased the volume and therefore reverberation time) was out of the question as it would have drastically changed the image of the building. One of the first outstanding systems for open spaces designed to simulate the acoustics of enclosures was installed by C. Jaffe for a series of performances by the St. Louis Symphonic Orchestra in a number of squares and parks around the city.

Despite a certain resistance to these systems of artificial reverberation at first (mainly by the music community, rather than by the public), today they are universally accepted and even a clear change of trend is envisaged for the coming century with the aid of Digital Sound Processing.

A third milestone in technological progresses is the appearance of computers in the 1960s and the spectacular breakthroughs witnessed subsequently. Since the 1980s powerful computer programs have been available, capable of simulating the behaviour of the sound field inside a room with a high degree of precision, which is due to their ability to perform a large number of different operations in a short time. These simulation programs have gradually been replacing the use of scale models, the purpose of which is to aid the architect in making decisions about the shape of the interior space and the surface materials, prior to building. Nowadays any attempt at determining the acoustic behaviour of a room without the aid of these programs and the computers that support them is inconceivable. Furthermore, in the last decade these simulation programs have been complemented by others that allow virtual sound generation. This has led to what is known today as Auralization, the name given to a process that simulates the hearing (preferably through headphones) of speech or a piece of music in a certain room, even before it is built.

It is from this point of view that we have approached the field of Computational Architectural Acoustics. This book consists of a significant amount of current knowledge on room acoustics, and has therefore been very fittingly placed in a series on Advances in Architecture.

It has been decided to present the contents in six chapters, written by authors or work teams, all of whom are researchers in this scientific-artistic-technical field. The first two chapters centre on the most outstanding aspects of room acoustics that have been studied in depth this century: absorption, sound reflection and diffusion, echo and reverberation. The next two chapters present studies of simulation models of the binaural experience of listeners in a room. Their main aim is the analysis of so-called subjective attributes of sound fields. Much of the current research on room acoustics is dedicated to perfecting these models. Finally, the last two chapters comprise work and research carried out in recent years by the authors on acoustics in very specific places: Concert Halls and Auditoria, on the one hand, and Churches, on the other; spaces which, though very different in kind, share the common factor of having been historically "the places of music".

September 1999
Juan J. Sendra

Chapter 1

Reverberation and diffusion

M.R. Schroeder
University of Göttingen, Germany
ATT Bell Laboratories (retired)

Abstract

Reverberation and diffusion are the reigning fundamental concepts in architectural acoustics. Reverberation describes the temporal aspects of sound transmission in a hall and diffusion captures its spatial qualities. Proper reverberation makes music sound smooth and pleasant. Sufficient diffusion gives the listeners the feeling of being "immersed" in the sound. But as in the Theory of Relativity, space and time are intimately intertwined. Sound diffusion affects the reverberation process and reverberation influences perceived spatialness. This chapter reviews the fundamental facts of reverberation and describes methods for improving sound diffusion.

1 Introduction

Room acoustics, and especially concert hall acoustics, is a subject that belongs at the intersection of physical science, engineering and art. The science of acoustics – how sound waves are propagated and reflected – is the foundation of room acoustics. Mechanical and electrical engineering govern the proper use of sound absorbing materials, public-address systems and artificial reverberation. Finally, the ability of people to hear and to differentiate different sounds is the basis of the artistic appreciation of speech and music.

Personal preferences can be quantified by modern methods of multidimensional scaling, but to satisfy disparate musical tastes is a difficult task. This challenge is further complicated by the widespread desire to build multipurpose halls that function well for lectures, dramatic theatre, intimate musical ensembles and large orchestras.

This chapter attempts to illuminate room acoustics from the following viewpoints:

- sound rays, echoes and reverberation
- chaotic interference of sound waves
- sound enhancement and artificial reverberation
- subjective preferences
- sound diffusion.

2 Sound rays, echoes and reverberation

In general, the wave equation cannot be solved for any realistic room shapes. Scientists therefore approximate the propagation of sound waves by sound rays. The *ray approximation* works particularly well for wavelengths small compared to the "obstacles" considered. Sound rays are indispensable for analyzing echoes and studying the wall-hugging "Whispering gallery" effect (see Figure 1).

Reverberation theories, too, are based on sound rays. Each time a sound ray hits an absorbing wall its energy is reduced by a factor $(1-\alpha)$, where α is the absorption coefficient. Given that the expected rate of wall collisions equals $cS/4V$, where S is the absorber surface area, the energy as a function of time t is given by an exponential decay:

$$E(t) = E_0 (1-\alpha)^{\frac{cS}{4V}t}$$

or, with the absorption exponent $a = -\log(1-\alpha)$,

$$E(t) = E_0 \exp(-a\frac{cS}{4V}t)$$

Reverberation time T is traditionally defined by a 60 dB decay, i.e. the ratio $E(T)/E_0 = 10^{-6}$. Thus we obtain

$$T = 55.3\frac{V}{acS} \tag{1}$$

called Eyring's reverberation time formula. Approximating a by α gives the original Sabine formula.

Both formulas neglect the shape of the enclosure and the placement of the absorber which can have substantial effects. Ignoring higher moments of the ray statistics and working only with the mean collision rate $cS/4V$ (or the mean free path $4V/S$) leads to additional errors. The proper way to calculate reverberation requires the solution of an integral equation.

Figure 1. Sound propagation in a "whispering gallery": the sound rays "cling" to the concave wall.

Traditionally, reverberation times were measured by exciting the enclosure with a brief burst of sound energy (pistol shots, noise bursts or tone pulses). The subsequent sound decay was then recorded on an oscilloscope or plotter and evaluated by curve fitting.

The statistical uncertainties of the noise excitation can be avoided by backward integration ("Schroeder integration") of the sound decay. Equipment overload is circumvented by the use of number-theoretic binary maximum-length ("Galois") sequences as an excitation signal and subsequent deconvolution on a computer.

Echoes and reverberation can be controlled by the proper placement of the right sound absorber. Narrow frequency ranges are best absorbed by acoustic cavities called Helmholtz resonators. Wideband absorption is achieved by lossy materials, such as mineral wool, either fully exposed or behind perforated panels. Absorption coefficients are measured in reverberation chambers using Eqn (1). Even *transparent* sound absorbers (for the German *Bundestag* in Bonn) have been realized by drilling micropores into plexiglass. In this manner the desired "fish-bowl" architecture of the building could be maintained while solving its cocktail of acoustic problems.

3 Chaotic interference of sound waves

The relation between reverberation time T and the bandwidth B of a resonance is

$$B = \frac{2.2}{T}$$

where the constant 2.2 equals $(3/\pi)\log_e 10$. With a reverberation time of 1.7 seconds, typical for mid-sized to large halls, the mode bandwidth therefore equals 1.3 Hz. Recalling the formula for the average mode spacing $c^3/4\pi V f^2$, we see that for frequencies above

$$f = \frac{1}{2\pi}\sqrt{\frac{3c^3 T}{\log_e 10 V}}$$

the average mode spacing will be smaller than the mode bandwidth and, for $f > f_c$,

$$f_c = \frac{1}{2\pi}\sqrt{\frac{c^3 T}{\log_e 10 V}} = 2000\sqrt{\frac{T}{V}}$$

on average three or more modes overlap, leading to a "statistical" response of the enclosure. Here T is measured in seconds and V in m^3.

More specifically, for frequencies above the "Schroeder frequency" f_c the complex sound transmission function of a reverberant enclosure (with negligible direct sound) will, as a function of frequency, approach a complex Gaussian process in the frequency domain with an exponential power spectrum (if the reverberation is exponential). All other characteristics of the sound transmission in "large rooms" follow from this observation. For f_c to fall near the lower end of the audio band (100 Hz, say) and $T = 1.7$ s, the volume V must exceed a modest 235 m^3 to guarantee good modal overlap. Thus, the statistical theory applies even to small enclosures in the entire audio range.

The theory predicts an average spacing of $4/T$ for the response maxima, which was once considered an important objective measure of acoustic quality. The standard deviation of the statistical responses is about 6 dB, independent of reverberation time or volume. Ironically, these and other objective measures were intended to *supplement* reverberation time (which was known to be insufficient as a predictor of concert hall quality). Yet they are either numerical constants or depend only on reverberation time.

If a public-address system is operated in a reverberant hall, high peaks in the statistical response can lead to audible acoustic feedback ("howling"). Inserting a frequency shifter with a small frequency shift (about 5 Hz) in the feedback loop will increase the margin of acoustic stability by several decibels.

If the Schroeder frequency f_c is expressed as a wavelength $\lambda_c = c/f_c$ and the Sabine value substituted for the reverberation time, then

$$\lambda_c = \frac{\pi}{3}\sqrt{\frac{A}{6}}$$

where $A = \alpha S$ expresses the total absorption by an equivalent "open window" area. This formula is independent of the units used!

Another important room acoustical parameter is the distance r_C from an omnidirectional sound source at which direct and reverberant sound energy densities are equal:

$$r_c = \frac{1}{4}\sqrt{\frac{A}{\pi}}$$

It was only recently discovered that there is a close relationship between r_C and λ_C ($r_C \approx 0.35\ \lambda_C$ for three-dimensional enclosures and $r_C \approx 0.16\ \lambda_C$ for two-dimensional spaces).

4 Sound enhancement and artificial reverberation

For optimum enjoyment music requires a proper portion of reverberation. Think of organ music composed for a cavernous cathedral, with a reverberation time of 4s. Or consider the romantic repertoire which sounds best with a reverberation time of 2s, with a rise toward low frequencies for the much desired "warmth". Not infrequently, natural reverberation is scarce or completely lacking, such as when the New York Philharmonic plays in Central Park. While perhaps tolerable to the aficionados stretched out on the Meadow, the lack of reverberation is jarring when listening to such a concert over the radio. Here, as in multipurpose halls, in recording studios and in electronic music, *artificial* reverberation is called for. A prime example is the 6000-seat Palace of Congresses in the Kremlin, designed primarily for political events, i.e. speech(es), for which a small reverberation time is optimum. But when the Bolsoi Theatre ran out of seating space, the Congress Hall had to take on opera as well and the required reverberation was "manufactured" in reverberation chambers in the subbasement and piped into the hall via loudspeakers.

Another favorite method of creating artificial reverberation were large steel plates or springs, but they introduced a "metallic twang". Simple feedback around a delay line also creates reverberation but it discolors the sound because of its comb-like frequency response ("combfilter"). Finally, in the late 1950s, the proper solution, "colorless" artificial reverberation, was found: *all-pass* reverberators whose complex transmission functions have a frequency independent magnitude. In the simplest case, an allpass reverberator can be realized by adding a negative-amplitude undelayed impulse to the output of a feedback-delay reverberator. Electronic allpass reverberators are now widely used, even in home entertainment. Digital reverberation networks using allpass and combfilters can even be designed to simulate sound transmission in concert halls, both existing and planned. This technique allows the pretesting of new designs before construction begins, thereby avoiding expensive architectural blunders.

An important application of artificial reverberation is multipurpose halls. Typically these are designed for high speech intelligibility, which means short reverberation times, and the reverberation required for music is added "electro-acoustically", that is to say through loudspeakers, as in the Palace of Congress mentioned above.

Intelligibility can also be enhanced electro-acoustically by "negative" reverberation, i.e. by the provision of extra direct sound – in other words *public-address systems* as used in many lecture halls and churches. In such systems, a multitude of loudspeakers, preferably loudspeakers *columns*, project the speaker's amplified voice directly into the audience.

For optimum intelligibility, the bass control should be turned down as far as possible because the low frequencies, which are not as effectively absorbed by hair and clothing as the higher frequencies impede rather than increase intelligibility. This effect is the result of "upward spread of masking" in the inner ear of humans, i.e. low frequencies mask the higher frequencies which carry most the speech information.

To maintain the illusion that the sound comes from the speaker's lips, sophisticated systems exploit the "precedence" or *Haas effect* by delaying the amplified sound enough to arrive at the listeners' ears *after* the "natural sound". Such a system was first successfully installed and operated in St. Paul's Cathedral in London.

5 Subjective preferences

What kind of acoustics do people actually prefer when listening to, say, classical music? The literature abounds with the results of subjective studies – some of a questionable character. Typically, trained (or naive) listeners have to rate the hall according to various categories such as warmth, brilliance, clarity and a dozen more – on a scale from 1 to 5, say. The subjective preferences scores are then averaged end correlated with the physical characteristics of the enclosures.

A better approach, however, is to abstain from such semantically loaded terms, which may mean different things to different people. Instead, the best approach is to simply ask listeners for each pair of concert halls which one they prefer. In order to make such comparisons possible, a selected piece of music is recorded by an orchestra in a reverberation-free environment, reproduced in the halls under investigation and recorded with stereo-microphones embedded in an artificial head.

Such recordings can be processed to recreate, at a listener's ears listening to loudspeakers in an anechoic environment, the original sound signals. Thus, listeners can instantly switch themselves from one hall to another and, on the basis of identical musical input, make reliable judgments.

The resulting preference scores are evaluated by multidimensional scaling which result in a *preference space*, typically of two or three dimensions. The first dimension, which may account for some 50% of the total variance, typically represents a "consensus preference" among the listeners, while the second dimension reflects individual differences in musical taste.

When the most significant preference dimension is correlated with the objective parameters, such as reverberation time or width of the hall, it is found that the high-and-narrow halls of yore, such as the Vienna Grosser Musikvereinssaal, are much preferred over the low-ceiling fanshaped halls of more modern design. Listeners also prefer "stereo" sound as opposed to the monophonic signals that are created by sound waves arriving from frontal directions. These two preferences (for narrow halls and for small "inter-aural correlation") are actually related: high-and-narrow halls deliver a preponderance of *lateral* sound, giving rise to a feeling of being "bathed" in sound as opposed to a feeling of detachment. In fact, lateral sounds seem to be the main reason for the observed preference of older halls.

6 Sound diffusion

Unfortunately, wide halls with a low ceiling are here to stay, enforced by economic dictates: wider halls mean more seats to sell and lower ceilings engender lower building costs (the air our ancestors needed to breath now comes from air conditioning rather than the extra air volume of high halls).

But can we recover the old acoustic advantages of high-and-wide halls? Leaving out the ceiling, and thereby eliminating frontal sound from overhead, might be helpful for the acoustics but is of course unacceptable in most climes. How about "diffusing" the ceiling reflection laterally? This can indeed be done by turning the ceiling into a reflection phase grating based on number-theoretic principles (see Figure 2). Making the depths of the troughs proportional to the quadratic residues of successive integers modulo a prime number p, say $p = 17$, such ceilings can be made to scatter sound into wide lateral angles over four musical octaves (see Figure 3). The quadratic residues form a periodic sequence, which for $p = 17$ looks as follows: 1, 4, 9, 16, 8, 2, 15, 13, 13, 15, 2, 8, 16, 9, 4, 1, 0; 1, 4, 9 etc. Such number-theoretic diffusors (called *Schroeder diffusors*) are now available commercially for installation in recording studios, lecture halls, churches, and living rooms, as well as concert halls.

Other diffusors are based on primitive elements in finite fields and the number-theoretic logarithm. At low frequencies such diffusors exhibit some sound desorption.

7 Conclusion

While the proper design of halls for music, opera, drama and lectures remains a challenging problem, especially if several of these purposes are to be combined ("multi-purpose halls"), modern methods of realistic simulation and accurate calculation should ease the design task. With increasing reliability of digital equipment and better transducers (loudspeakers and microphones) electroacoustic means for improving and modifying room acoustics should become widely acceptable.

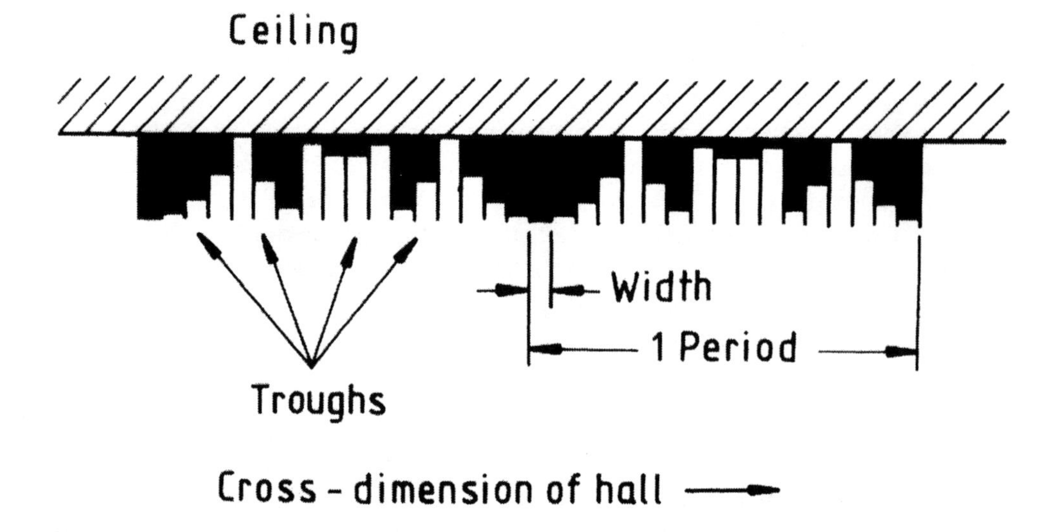

Figure 2. Number-theoretic reflection phase grating based on succesive quadratic residues modulo the prime number 17. The pattern repeats with a period length of 17 and scatters frequencies over a range of 1:16, corresponding to four musical octaves.

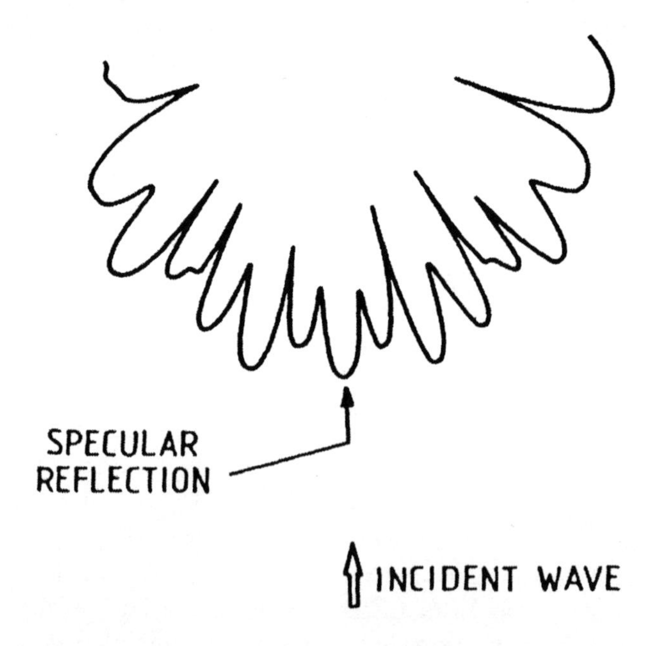

Figure 3. Reflection pattern from the phase grating shown in figure 2 for vertically incident sound. The wide angular scatter of sound energy also obtains for oblique incidence.

Acknowledgement

The chapter was adapted from an article by the author in the *Encyclopaedia of Electrical and Electronics Engineering* (John G. Webster, editor) published by John Wiley and Sons, Inc.

References

[1] Ando, Y., *Concert Hall Acoustic*, Springer-Verlag, Berlin, 1985.

[2] Ando, Y. and Nelson, D., *Music and Concert Hall Acoustics*, Academic Press, San Diego, 1997.

[3] Beranek, L., *Concert and Opera Halls: How They Sound*, Acoust. Soc. Amer., Woodbury, New York, 1996.

[4] Blaubert, J., *Spatial Hearing, The Psychophysics of Human Sound Localisation*, MIT Press, Cambridge, MA, 1983.

[5] Kuttruff, H., *Room Acoustics*, Applied Science Publishers, Barking, Essex, UK, 1973.

[6] Cavanaugh, W.J. and Wheterill, E.A., *Walle Clement Sabine Centennial Symposium*, Acoustic. Soc. Amer., Woodbury, New York, 1994.

[7] Pierce, A.D., *Acoustics: An Introduction to Its Physical Principles and Applications*, McGraw–Hill Book Co., New York, 1981.

[8] Sabine, W.C., *Collected Works on Acoustics*, Peninsula Publishing, Los Altos, California, 1992.

[9] Schroeder, M.R., *Number Theory in Science and Communication, Third Edition*, Springer-Verlag, Berlin, 1997.

[10] Tohyahma, M., Suzuki, H. and Ando, Y., *The Nature and Technology of Acoustic Spaces*, Academic Press, San Diego, 1995.

Chapter 2

Sound absorbing materials and sound absorbers in enclosures

C. Díaz and A. Pedrero
Laboratorio de Acústica y Vibraciones. E.T.S. de Arquitectura, U.P.M., Avda Juan de Herrera 4, 28040 Madrid, España
Email: jdiaz@aq.upm.es

Abstract

Sound absorption in an enclosure has a great effect upon its sound field and affects in a very significant way its acoustic parameters. In order to control the sound field in an enclosed area according to its use, sound absorbing materials and devices are usually employed. This chapter characterizes the behavior of the most common sound absorbing materials and devices, and indicates some of the methods for measuring sound absorption coefficients.

1 Introduction

Sound absorption is a physical phenomenon resulting from the dissipation of sound energy into heat energy. In an enclosure this dissipation can be produced by the propagation of sound waves in air and their incidence on the surfaces of the enclosure (walls, ceiling, floor) or on the people and objects situated in its interior.

The quantity of sound absorption in an enclosure is a very important factor in its sound quality that decisively influences the phenomena related to the reverberation and the distribution of the sound pressure level in the enclosure´s reverberant field.

Sound absorbing materials and devices are used to control the sound field in an enclosure according to its use, and therefore it is important to know their absorption characteristics in relation to the frequency. In reality, they are passive sound absorbers.

2 Absorption of sound and absorption coefficients

When a sound source emits in an enclosure, the level of sound pressure in the reverberant field and the reverberation time are affected by the absorption of sound energy by the surfaces of the confined space as well as by people and objects inside the area.

The ability of a material to absorb sound energy is characterized by the sound absorption coefficient. This is a dimensionless quantity that is defined as the ratio between the absorbed (not reflected) and the incident acoustical energies on the material. It is written as follows:

$$\alpha = \frac{E_a}{E_i} \tag{1}$$

α is the sound absorption coefficient; E_a and E_i are, respectively, the absorbed and incident acoustical energies on the material. This coefficient depends on the frequency and the angle of incidence of the sound waves.

Observed from the interior of an enclosure, absorbed sound energy is the sum of energy dissipated among surfaces, people and furnishings E_d, and the energy that is transmitted to the exterior, E_t. Figure 1 shows a drawing of this process.

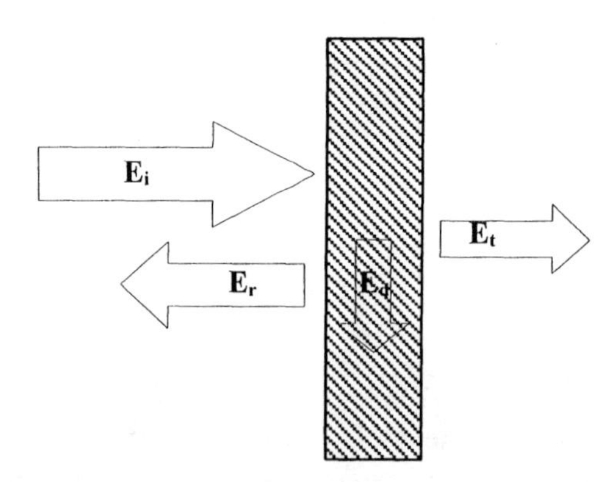

Figure 1. Drawing of the energetic relations on the enclosure wall.

The above mentioned definition of the sound absorbing coefficient suggests several possible ways of quantification since the fraction of incident energy absorbed by a surface depends on the angle of incidence. Out of all these possibilities, the following absorption coefficients are noteworthy for their theoretical importance and their applications.

2.1 Statistical or random sound absorption coefficient, α_{ran}

This coefficient is defined, for an infinite size plane surface, as the ratio between the sound energies absorbed and incident on the surface, when the sound incident field is perfectly diffuse. This means the incident angle varies constantly in a random way. The maximum theoretical value of this absorption coefficient for a plane surface is, $\alpha_{ran} \cong 0.96$, e.g. Morse and Ingard [1].

2.2 Sound absorption coefficient in reverberant chamber or Sabine sound absorption coefficient, α_S

This coefficient is related to acoustical reverberation in an enclosure. Most of the coefficients that are published are obtained by the reverberant chamber method. The manufacturers of acoustic absorbing materials usually give the Sabine absorption coefficient in six octave bands with frequencies centered on 125, 250, 500, 1000, 2000 and 4000 Hz. The Sabine absorption coefficient is higher than the random absorption coefficient. In some of the frequency bands, values that are greater than a unit may be obtained. This difference is due to the diffraction phenomenon of the incident acoustic field on the test material and the fact that the sound field in the enclosure is not completely diffuse.

The Noise Reduction Coefficient (NRC) of an acoustic absorbing material is the average value of the absorption coefficients α_S of the material with the octave bands centered on the frequencies of 250, 500, 1000 and 2000 Hz; this average is rounded off to the nearest multiple of 0.05. Two materials may have identical NRCs but very different absorption characteristics.

2.3 Sound absorbing coefficient for plane waves and normal incidence α_n

This coefficient is directly related to the characteristic acoustic impedance of the material and its value is determinated by the impedance tube method.

In the case of materials known as "local reaction", and under certain mounting conditions, the Sabine coefficient and normal incidence are correlated.

3 Sound absorption in enclosures

The sound absorption of a surface area S is obtained by multiplying its sound absorption coefficient by its area. The total sound absorption of all the surfaces of the enclosure can be calculated by the equation:

$$A_{\text{surfaces}} = \sum_{i=1}^{i=n} \alpha_i . S_i \quad \text{m}^2 \tag{2}$$

The sum includes all the interior surfaces S_i with different absorption coefficients α_i. The sound absorption is measured in m^2.

The average sound absorption coefficient of an enclosure, α , is defined as the ratio between the total sound absorption and the total area of the enclosure.

Total sound absorption in a closed area is the sum of the sound absorption produced by the interior surfaces, $A_{surfaces}$; furnishings, $A_{furnishings}$; air, A_{air} ; and people situated in its interior, A_{people}:

$$A = A_{surfaces} + A_{furnishings} + A_{air} + A_{people} \tag{3}$$

Separate objects such as people, furniture, special absorbers, etc. cannot be considered plane surfaces and therefore, the concept of equivalent sound absorption area is used. This is the imaginary area of a sound absorber with a unit absorption coefficient that produces the same total sound absorption.

3.1 Sound absorption by air

When an acoustic disturbance propagates in the interior of an enclosure, the density of sound energy diminishes due to viscosity degradation and molecular relaxation processes, which equally affect all the points in the enclosure volume.

At the frequencies that are of interest to enclosure acoustics, the amount of sound absorbed by air is not significant in small size enclosures. However, in those of large dimensions, air absorption can be important at frequencies higher than 2000 Hz. Air sound absorption in a large enclosure is obtained by means of the formula:

$$A_{air} = 4mV \quad \mathrm{m}^2 \tag{4}$$

m is the air attenuation coefficient per meter and V is the volume of the enclosure. Table 1 shows some values of m, at 20°C with normal atmospheric pressure and several relative humidity values [2].

Table 1. Air attenuation coefficient m at 20°C and normal atmospheric pressure, in $10^{-3}\mathrm{m}^{-1}$.

Relative humidity	Frequency center octave band, Hz		
	1000	2000	4000
50 %	1.05	2.22	6.55
60 %	1.10	2.12	5.85
70 %	1.12	2.02	5.20

3.2 Sound absorption by people

In enclosures such as classrooms, libraries, churches, auditoriums, etc. sound absorption due to people situated in the interior represents the largest contribution to sound absorption. One method for calculating audience sound absorption consists of multiplying the given sound absorption area equivalent per person in a certain situation, by the anticipated number of people in the enclosure.

In large enclosures, according to results obtained by Beranek [3] and Kosten [4], when a great number of people are put together on a surface, this produces a very absorbent area and therefore the sound absorption due to the audience is not obtained by multiplying the sound absorption area equivalent per person by the number of people. It is more realistic to multiply the "acoustic area of enclosure" by some sound absorption coefficient equivalents obtained experimentally. Table 2 shows the latest results obtained by Beranek [5].

Table 2. Absorption coefficients in large enclosures.

Seats, unoccupied						
Frequency, Hz	125	250	500	1000	2000	4000
Heavily upholstered chairs	0.72	0.79	0.83	0.84	0.83	0.79
Medium upholstered chairs	0.56	0.64	0.70	0.72	0.68	0.62
Lightly upholstered chairs	0.35	0.45	0.57	0.61	0.59	0.55
Occupied audience, orchestra and chorus						
Frequency, Hz	125	250	500	1000	2000	4000
Heavily upholstered chairs	0.76	0.83	0.88	0.91	0.91	0.89
Medium upholstered chairs	0.68	0.75	0.82	0.85	0.86	0.86
Lightly upholstered chairs	0.56	0.68	0.79	0.83	0.86	0.86

4 Relationship between sound absorption and acoustic parameters in the enclosure

In an enclosure, sound travels short distances before it strikes interior surfaces or other obstacles that reflect and/or absorb the sound in a way that decisively influences its acoustic properties.

An enclosure is an active element that affects a signal emitted by a sound source in its interior since it can modify its distribution by means of reflections that can produce echoes. Furthermore, it can modify the signal′s spectrum, increase the sound pressure level, modify the listening conditions at different points in the enclosure, etc.

4.1 Sound absorption and reverberation time

Reverberation in an enclosure is referred to as the process of sound energy persistence and decrease within the confined area after the sound source has stopped its emission. This process is produced by multiple sound wave reflections from the surfaces and objects in the enclosure. W.C. Sabine defined enclosure reverberation time, T_{60}, as the time interval taken for the sound pressure level to decay by 60 dB after the source stops. In traditional acoustic formulation, reverberation time is considered the main parameter in determining the acoustic quality of an enclosure according to its use.

The oldest theories on sound reverberation in a closed area which led to the Sabine reverberation time equation [6–9] came from the hypothesis that the sound field is diffuse, which means that sound energy is the same at all points and in all directions of the enclosure.

According to Sabine, the equation for calculating reverberation time, T_{60}, is as follows:

$$T_{60} = \frac{0.16V}{A} \quad \text{s} \tag{5}$$

in which V is the enclosure volume in m^3 and A represents the total sound absorption in the enclosure in m^2. The reverberation time, according to the above equation, does not depend on the shape of the closed space or on how the sound absorbing materials and devices are distributed. Other equations for reverberation time were later developed [10–13], and they continued to consider the sound field in an enclosure as being diffuse.

In all reverberation time equations the same conclusion is reached; as sound absorption increases, the reverberation time decreases. Consequently, the enclosure′s acoustic response can be controlled by appropriately selecting the sound absorption materials and devices that are placed in the enclosure.

In 1969, Jordan [14] introduced the new parameter *Early Decay Time*, *EDT*. It is the time interval that sound takes to decrease from 0 to –10 dB, from the instant a sound source has stopped emitting, multiplied by 6. It is calculated by the same process as reverberation time. This parameter is important when the acoustic qualities of enclosures are compared with one another, and also the subjective judgments of the reverberation in them are correlated better than with the classical reverberation time.

The sound absorption coefficients of materials and devices that usually appear in books, manuals, and manufacturers′ catalogues are obtained by the reverberation chamber method, which employs the Sabine equation for reverberation time. It is important to keep this in mind when using other formulae to calculate the reverberation times that are different from Sabine′s. For example, if the sound absorption coefficient of a material obtained in a reverberant chamber is α_S, its relation to the sound absorption in Eyring′s formula is

$$\alpha_E = 1 - e^{-\alpha_S} \tag{6}$$

At present, numerous studies have demonstrated that other parameters also characterize acoustic quality in an enclosure and have analyzed in a more detailed manner the temporal and spatial distribution of the sound field within the enclosure. All these parameters are based on the analysis of the enclosure′s impulse response, by establishing energy relations between the different parts of this response.

The impulse response in an enclosure describes sound energy as a function of time that is obtained at a specific point in the enclosure, and produced by an impulsive sound disturbance. In a closed area, direct sound emitted by a source can reach any listener in a time interval from 20 to 200 ms, depending on the distance between the sound source and the listener. Shortly after, the same sound, which was emitted by the source, is received from various reflecting surfaces, namely walls and the ceiling. These reflections arrive after certain time delays t_1, t_2, ...,t_n, in relation to the direct sound. This first group of reflections that reaches the listener in a time interval of approximately 80 – 100 ms after the direct sound is usually known as early reflections. The reflected sound that arrives with a greater delay than the aforementioned sound, does so in a continuous and rapid manner from all directions, merging into the reverberant sound. Figure 2 shows an impulse response or reflectogram.

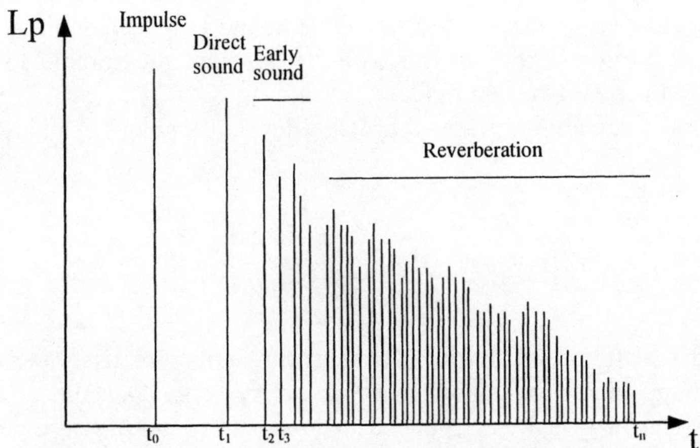

Figure 2. Direct and reflected sound impulse response between a sound source and a listener.

Unlike statistical methods used to obtain reverberation time T_{60}, impulse response is closely related to methods suggested for geometrical acoustics: it is based on the hypothesis that sound energy emitted by a source is distributed in a finite number of rays that are propagated in the interior of the enclosure. When a ray reaches an interior surface of the enclosure, it is reflected from it, whereby it loses part of its energy due to sound absorption by the surface. Sound energy at a specific point in the enclosure is calculated as the sum of energy from the different rays that converge on that point at a certain instant of time.

From the above, it can be deduced that impulse response is not an attribute of the enclosure, as it was in the case of reverberation time. Impulse response depends, in each case, on the position of the source and the receptor, the directivity factor of the source, the geometry of the enclosure, as well as the placement of sound absorption materials and devices in the interior. Based on these facts, there are surfaces that have a greater probability of receiving rays than others. Therefore, the effectiveness of a sound absorption material not only depends on its sound absorption coefficient, but also on the probability of sound waves reaching the surface of the material. Some simulation programs of acoustic in an enclosure, take into account this probability of incidence.

4.2 Sound absorption and the sound pressure level in the enclosure

When a source emits sound energy in an enclosure in a permanent manner, the sound pressure level at each point in the enclosure can be considered a consequence of two overlapping sound fields, the direct and reverberant sound fields.

The direct sound field is formed by energy that reaches the point in question directly from the source, without being affected by any reflection. The reverberant sound field is produced by multiple sound reflections from the surfaces in the enclosure and is considered diffuse. The relative contribution of each one of these fields on the total sound pressure level which exist at each point of the enclosure, basically depends on the distance between the sound source and the point in question. For positions that are very near the source, direct sound is the predominating factor, whereas for those positions that are far away, the sound pressure level is almost exclusively due to the reverberation field.

The formula for calculating the sound pressure level at a point in the interior of an enclosure is

$$L_P = L_W + 10\log\left[\frac{Q}{4\pi r^2} + \frac{4}{R}\right] \quad \text{dB} \tag{7}$$

in which L_W is the sound power level of the source, Q is its directivity factor, r is the distance of the sound source to the point and R is the enclosure constant.

The directivity factor Q is the ratio of the mean-square sound pressure produced by the sound source at distance r from an actual source, to the the mean-square sound pressure at the same distance from a non directional source radiating the same acoustic power. The directivity factor is a function of the direction, frequency and position of the source in the enclosure.

The enclosure constant R, is defined as

$$R = \frac{A}{1-\alpha} \quad \text{m}^2 \tag{8}$$

Given eqn. (7), it can be concluded that, by controlling the sound absorption of the materials present in an enclosure, the sound pressure level in the reverberant field is controlled. The greater the average sound absorption coefficient value of an enclosure is, the less the sound pressure level in the reverberant field will be.

The distance at which direct and reverberant sound field contributions are equal is called the enclosure reverberation radius and is expressed as

$$r_e = \sqrt{\frac{QR}{16\pi}} \quad \text{m} \tag{9}$$

The placement of sound absorption material in an enclosure only affects the reverberant field. In order to control the noise level in the direct field area, other measures must be taken.

The variation of sound pressure level in a reverberant field, due to an increase in sound absorption, can be calculated by the following formula:

$$\Delta L_{\mathrm{P}} = 10\log\left(\frac{R_1}{R_2}\right) \quad \mathrm{dB} \tag{10}$$

in which R_1 and R_2 are the enclosure constants, before and after the placement of more sound absorption in the enclosure.

As a point of fact, all enclosures by nature and use have a certain initial sound absorption. However, the surface available for applying additional absorption rarely exceeds a quarter of the total surface. Consequently, the maximum reduction of the sound pressure level due to increase in sound absorption in an enclosure is approximately 10 dB, although the usual reductions obtained range from 2 to 6 dB.

When the sound field in an enclosure it not diffuse, the decrease in the sound pressure level can be different from that which was previously calculated. For example, if the first sound reflections are produced from surfaces that have a sound absorption coefficient greater than the average absorption coefficient in the enclosure, the reverberant sound energy reflected by these surfaces to the enclosure will be small. Thus, it will produce an additional reduction in the sound pressure level in the reverberant field. This case occurs very often in non-acoustically conditioned enclosures when a very directive sound reinforcement system is employed. In this type of enclosure the area of greatest absorption is the audience, towards which loudspeakers are normally directed. In order to quantify this effect, some authors [15] have suggested the use of a modifier coefficient M_{a}, known as an architectural modifier, that results in a modified constant of the enclosure R', so that $R' = RM_{\mathrm{a}}$.. This modifier is expressed as

$$M_{\mathrm{a}} = \frac{1-\overline{\alpha}}{1-\overline{\overline{\alpha}}_{\mathrm{c}}} \tag{11}$$

in which $\overline{\alpha}$ is the average sound absorption of the enclosure and $\overline{\alpha_c}$ is the sound absorption coefficient of the incident surface.

The above expression of architectural modifier is valid in the case in which all the energy emitted by the sound source reaches the surface in question. This definition can be completed by taking into consideration the fact that energy emitted by a source usually does so in all directions. Thus, the complete equation for the architectural modifier is

$$M_a = \frac{1-\bar{\alpha}}{1-\bar{\alpha}_c}\frac{Q}{Q_0} \tag{12}$$

in which Q is the directivity factor of the source and Q_0 is the directivity factor of a theoretical source that emits all its energy in the direction of the incident surface. In eqn. (7), it would be necessary to substitute R for $R' = RM_a$.

4.3 Sound absorption coefficient and speech intelligibility

The sound phenomena that interfere with speech intelligibility are mainly background noise, reverberation and echoes.

Background noise masks speech, reducing its intelligibility. This phenomenon is more accentuated the closer the noise level is to the speech level which the listener perceives.

There are several indices which are used to estimate intelligibility as a function of the signal-to-noise ratio (*SIL* [16], *Articulation Index* [17], etc). As a general rule, a signal-to-noise ratio greater than 25 dB is considered to be the optimum condition for understanding speech.

Reverberation produces a blurring effect on the speech signal which makes the reverberant energy caused by the pronunciation of a phoneme to interfere with the direct sound energy from the phonemes that follow. This effect can be clearly observed in Figures 3 and 4, which show the waveforms and the spectrogram of two signals recorded by a microphone near the speaker. The second one is the same signal recorded at a distance of 3 m in an enclosure with moderate reverberation. It is observed that the energy of each phoneme is absolutely defined in the nearby position, whereas reverberation mixes the energies of the different phonemes in the position farther away.

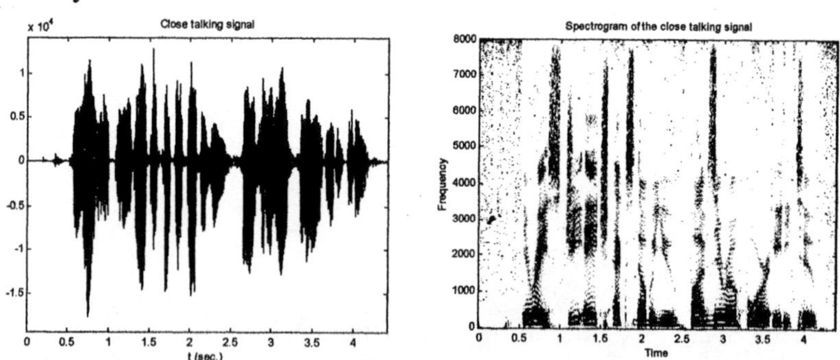

Figure 3. Spectrogram of the close talking signal.

The parameter that establishes the losses of intelligibility due to reverberation is the ratio of direct and reverberant energies that a listener receives. As a general rule, the greater this ratio is, the better the intelligibility. This effect can also be seen in several indices that estimate speech intelligibility, such as the so-called AL_{cons} [18], *STI* or its simplified version *RASTI* [19].

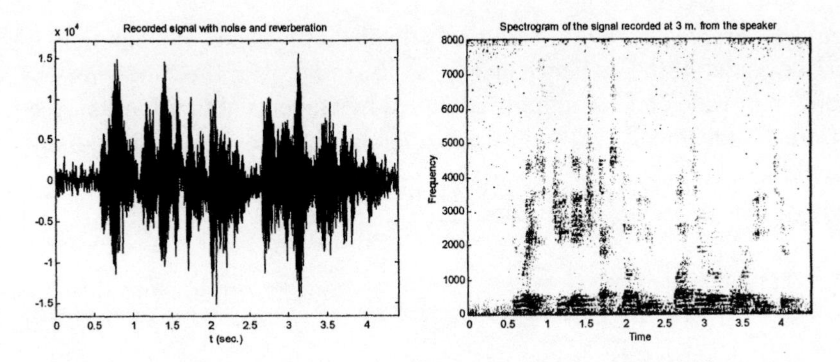

Figure 4. Spectrogram of the signal recorded at 3 m from the speaker.

Reflections can contribute positively to intelligibility. This is the case when they reach the listener in a delay time that is less than 50 ms with regard to the direct signal. In these cases, the early reflections reinforce the direct signal, improving its intelligibility. However, reflections with a delay greater than 50 ms are considered, as far as intelligibility is concerned, as if they were background noise. The loss that they produce depends on two factors, their energy and the delay time in relation to the direct signal.

As the previous information indicates, sound absorption in an enclosure plays a fundamental role in controlling intelligibility, since it conditions the phenomena that take part in its loss.

With high sound absorption we can reduce the noise level in a reverberant field, which means an important improvement in the signal-to-noise ratio. On the other hand, sound absorption in an enclosure conditions its reverberation whereby an appropriate selection of sound absorption materials and devices can provide an optimum intelligibility.

The placement of sound absorption materials and devices will condition the energy of each one of the reflections that reach the listener. In order to improve the intelligibility, the early reflections can be enhanced by placing reflective materials on the surfaces that produce them and by providing high sound absorption elements for the surfaces that generate delayed reflections.

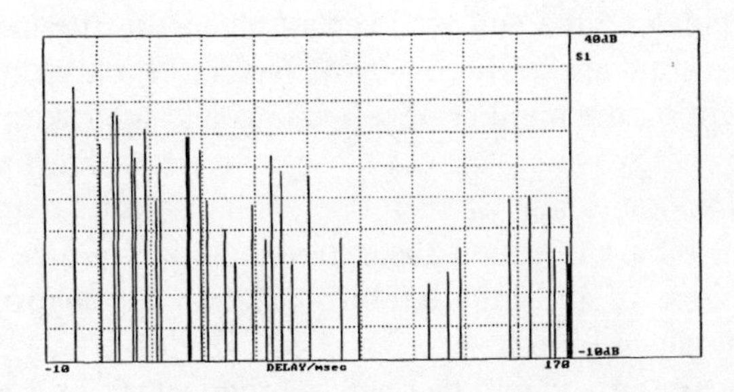

Figure 5. Reflectogram with reflecting walls and ceilings.

Figures 5–7 represent the reflectograms obtained by means of a simulation program taken in the center of the audience area of a classroom, with the following dimensions: 18 m long, 8 m wide and 3.5 m high, occupied by students at their desks. According to the different placements of sound absorption material, the reflectograms change considerably.

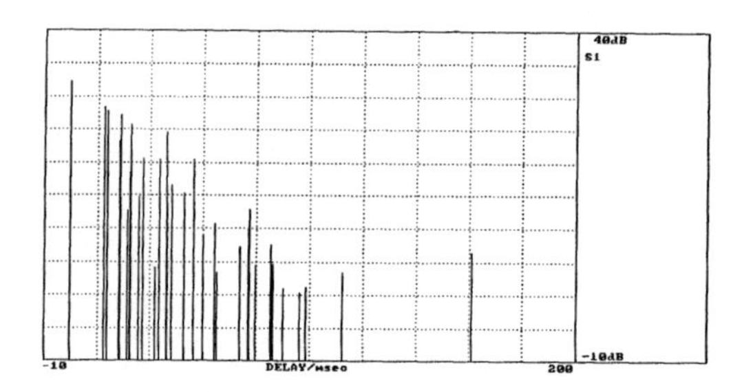

Figure 6. Reflectogram with sound absorption material on the back wall.

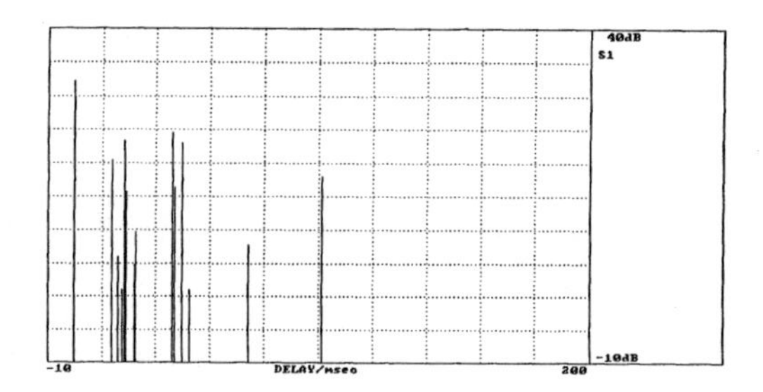

Figure 7. Reflectogram with sound absorption material on the ceiling.

4.4 Sound absorption and other acoustic parameters

Most of the parameters used to define the acoustic quality of an enclosure are based on detailed analysis of the enclosure impulse response. In some cases the energy ratios of the different impulse response areas are quantified, whereas in others the provenance of the direction of the reflections is also taken into consideration.

For every particular source and receiver arrangement, the impulse response of an enclosure depends fundamentally on two factors. Firstly, the geometry of the enclosure will condition the number of reflections and their temporal distribution. On the other hand, the energy which each of these reflections contributes is determined by the sound absorption coefficient of each one of the reflective surfaces. Therefore, by using a correct distribution of sound absorption materials and devices in an enclosure, the temporal envelope of the impulse response can be controlled, and hence, the parameter values obtained from it.

The most important parameters that can be obtained from the early-to-late energy ratio are:

The Clarity Index [20], C_{80}, which is the ratio, expressed in dB, between energy received in the first 80 ms and the energy received afterwards.

The Definition [21], D_{50}, which is the ratio between energy received in the first 50 ms and the total energy. This is commonly expressed as a percentage.

The Center Time [22], T_C, is the instant of time in which energy received before that instant is equal to that received afterwards, in the impulse response.

These parameters have optimum values when an important amount of initial energy exists. Thus, the present tendency in acoustic design is to foment the appearance of early reflections by means of an appropriate geometrical design, as well as by providing the surfaces which produce them, with materials that have a low sound absorption coefficient.

Early lateral reflections have also been proven to be important especially when evaluating the spaciousness of an enclosure. One criterion used for evaluating this factor is called *Lateral Efficiency* [23], *LE*. It is defined as the quotient of lateral energy in the time interval of 25 ms (some authors choose 5 ms) and 80 ms after the arrival of direct sound from a bidirectional microphone and the energy received in the first 80 ms by an omnidirectional microphone. In order to achieve appropriate lateral efficiency, it is necessary to foment the appearance of early lateral reflections.

Another important factor when evaluating acoustic quality in an enclosure is the sound amplification effect which the enclosure produces from the sound emitted by the source. The most characteristic parameter for evaluating this aspect is the denominated *Total Sound Level* or *Strength Index* [24], *G*. It is defined as the ratio, in dB, between the total energy in a certain position in the enclosure and the direct energy that would be received 10 m away in an anechoic chamber, for the same sound source operating at the same power level. It is obvious that the role of enclosure sound absorption is fundamental when evaluating this amplification. Although in this case it is not so important the temporal distribution of the energy supplied by the different reflections.

5 Sound absorbing materials and sound absorbers

The sound absorption coefficient of a material or an absorbing device, depends on the frequency, the incident angle of the sound, and varies considerably according to its mounting or installation. It is advisable to follow the manufacturer´s technical information in a correct manner. Different standard mountings exist for performing laboratory tests [25–27].

It is customary to classify the sound absorbing materials and sound absorbers according to sound energy degradation processes. This classification can be as follows:

Porous sound absorbers: their structure is granular or fibrous and the skeleton can be either rigid or elastic. The absorption of these materials is greatly influenced by the thickness of the material and the distance between it and the rigid surface behind.

Resonant sound absorbers: which can be simple or associated. The simple kind can, at the same time, subdivide into the Helmholtz type, membrane or panels type. The associated resonators can be in series and in parallel (usually with circular or grooved perforations).

Most of the commercially available sound absorbers are a combination of the two previously mentioned. Thus, they are called mixed sound absorbers.

There are sound absorbers with sound absorbing characteristics that vary gradually. They are known as anechoic absorbers.

Some sound absorbing materials are used to form continuous surfaces, by means of acoustic plasters. They are applied wet with a buttering trowel or spray-gun and are usually made of materials that are fibrous, organic or not, together with a binder.

In reality, the most widely used sound absorbing materials and devices are porous materials, absorbent panels and Helmholtz resonators or a combination of these.

5.1 Porous materials

They are made up of a solid, rigid or flexible structure, with irregular pores which are connected to each other and to the exterior. The sizes of the pores are much smaller than the length of the sound waves, generally less than 1 mm. Sound waves penetrate easily into the material and start vibrating the molecules in the air of the pore. The vibrations are absorbed by air viscosity and by the loss of heat energy in the structure of the material, hence decreasing the sound energy. At high frequencies the losses of sound energy increase due to friction.

Examples of absorbent porous materials are mineral and organic fibers, plastic foams, porous metals and ceramics, etc.

There are a series of physical parameters relevant to sound absorption by porous materials. These are porosity, flow resistivity, structure factor, tortuosity, mass density, compressibility, etc.

Porosity, *H*, is defined as the ratio between the volume occupied by the pores, in this case the air, and the total volume. Porosity can be expressed in terms of the air density, ρ_a , and the porous medium, ρ_m:

$$H = 1 - \frac{\rho_a}{\rho_m} \tag{13}$$

The porous materials that are normally used as absorbers have porosity values that range from 90 to 98%. There are several methods for measuring porosity [28].

The friction against the surface of fibers or particles that form the porous structure acts as an acoustic resistance whose value is a function of the resistance of the material to the direct flow of air. *Flow resistance* [29], *R*, is the ratio between the difference in pressure through a sample of material and the volume of air which it crosses. It is expressed as follows:

$$R = \frac{\Delta p}{q_V} \quad \text{Pa s/m}^3 \tag{14}$$

in which Δp is the difference in pressure, in Pa, between both sides of the sample and q_V is the volume of air which it crosses, in m^3/s.

Specific flow resistance, R_S, is defined as:

$$R_S = RS \quad \text{Pa s/m} \tag{15}$$

in which S is the area of the section of the sample in m^2.

The volume, q_V, is related to the linear velocity of the air flow by means of the formula: $u = q_V/S$, in which u is the linear velocity of the air flow in m/s, and S is the cross-section of the sample, in m^2.

The optimum value of flow resistance is a function of the frequency, thickness of the material and the mounting system. Its value must be between certain limits for absorption to be at its possible maximum. If friction is too high, the sound waves have difficulty penetrating, and if it is very low, friction is very small. This magnitude is not a constant of the material, since it depends on the thickness of the sample, d. An independent magnitude is air flow resistivity, r, which is the most important physical characteristic of a porous material. If the material is homogeneous, it is defined as

$$r = \frac{R_S}{d} = \frac{\Delta p}{ud} \quad \text{Pa s m}^{-2} \tag{16}$$

Concerning porous materials, air flow resistivity mainly depends on the density and size of the fiber. It can be considered independent of the air flow velocity only at low velocities; as these increase, r increases. However, a linear relation between them does not exist. In fibrous porous materials, the material fibers are generally situated parallel to the surface of the material. Flow resistivity in a perpendicular direction to the surface is greater than in a parallel direction [30, 31].

There are also materials with closed pores that are not interconnected with each other, or with the exterior, and are not good sound absorbers. Nevertheless, they are efficient thermal insulators.

From the point of view of sound absorption, there are differences between porous absorbing materials with a flexible structure and rigid structure. In the former case, with examples such as mineral wools, foams, porous cardboards, etc., the absorption increases with the frequency, whereas, in the latter case, with examples such as porous stones, microperforated sheets, etc., the absorption coefficient can vary significantly, especially at low frequencies.

Different authors have tried to establish models which would explain the behavior of these materials, but the laws are too complex. A review of the models used can be found in two works carried out by Attenborough [32, 33]. Some experimental parameters for general use are Delany and Bazley's [30] proposals.

Porous materials also absorb sound energy from the diffuse reflection produced by superficial layers of porous materials. The diffuse reflection is not affected by the thickness of the material. The surface finish of porous absorbing material can modify considerably its sound absorbent characteristics.

Porous materials are usually employed to control reverberant sound in an enclosure, to reduce the sound produced by pipes and to increase airborne insulation by inserting them as a filling between the walls.

5.1.1 Influence of porous material thickness

The thickness of a porous material greatly influences the improvement of sound absorption. When the material thickness increases, the sound absorption coefficient rises to lower frequencies. Figure 8 shows the effect on the sound absorption coefficient according to the variation in thickness of a porous material with a density of 15 kg/m^3 placed directly on a rigid wall. The tests were done in compliance with the ISO, R-354.

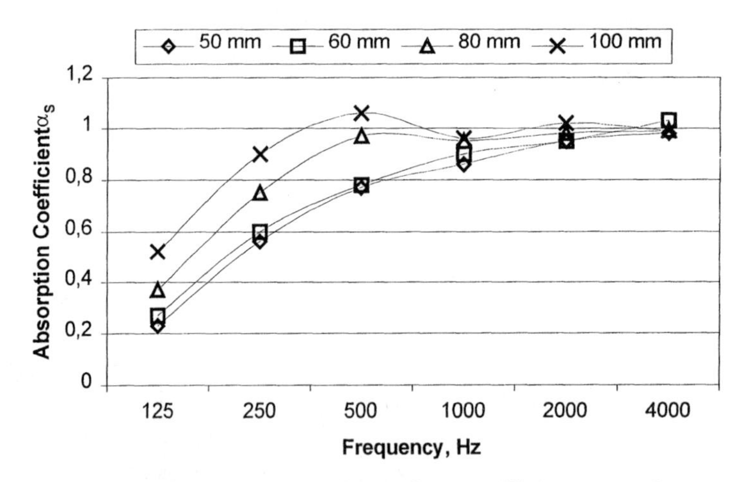

Figure 8. Relation between the sound absorption coefficient, α_S of a porous material and the material thickness.

5.1.2 Effect of an air space behind the absorbing material

The sound absorption coefficient of a porous material varies greatly according to how it is placed. The presence of an air space between the porous material and a rigid impermeable wall increases considerably the sound absorption at low frequencies.

Figure 9 shows this effect in the case of a fiberglass sound absorber, 25 mm thick, at different distances from a rigid impermeable wall.

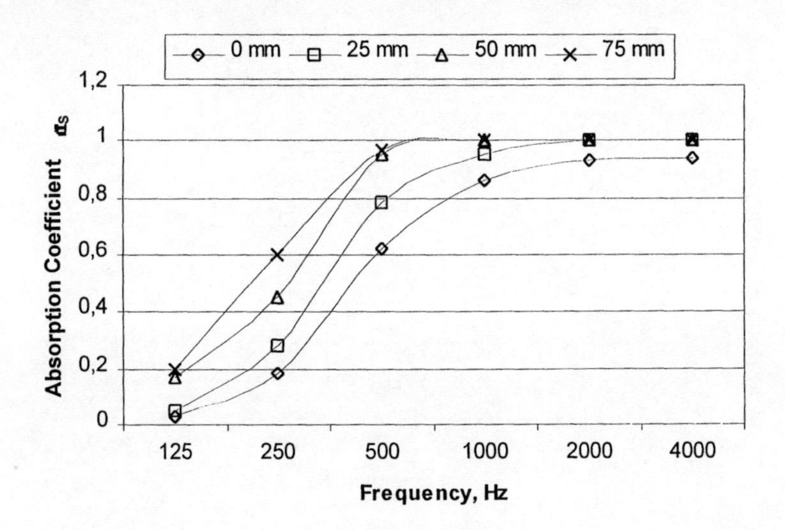

Figure 9. Effect of an air space behind an absorbent material.

5.1.3 Effect of the absorbent material density

Figure 10 demonstrates the small effect that the density of an absorbent material has on the sound absorption coefficient. The results have been obtained according to the standard ISO R-354, for a sample of rockwool, 5 cm thick with different densities, placed on a rigid impermeable surface.

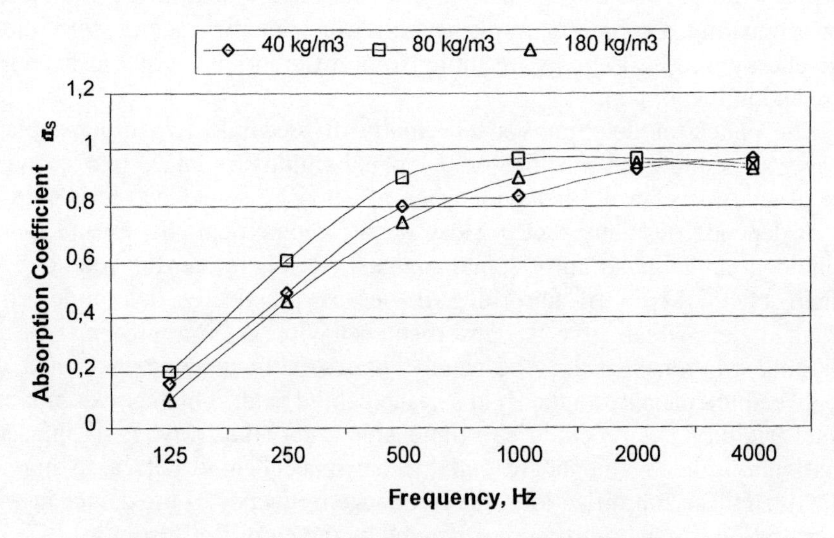

Figure 10. Effect of absorbent material density.

Figure 11 shows the absorption behavior at normal incidence of a porous material: silica, 8 cm thick, with and without a leather membrane cover.

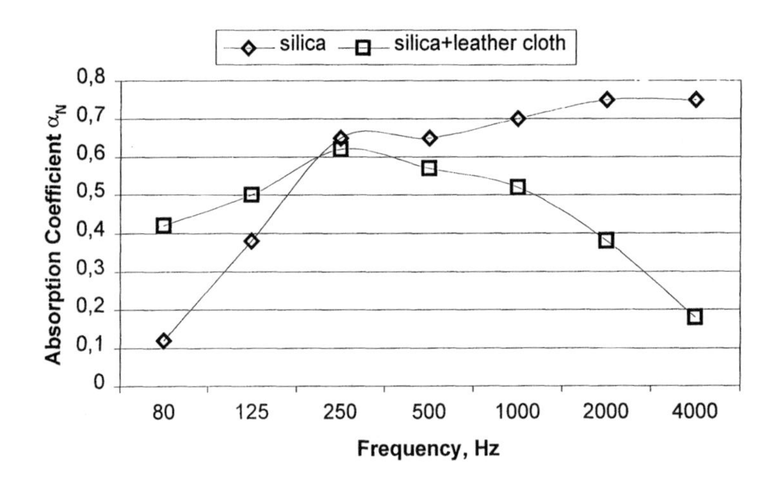

Figure 11. Effect of a membrane on a sound absorbing material.

5.2 Acoustical panels

Acoustical panels are basically made up of an airtight material fixed at a distance from a rigid surface, leaving an airtight gap between both. A typical example of these devices is a board made of plywood, nailed or glued to a frame, at a distance from a rigid surface.

This kind of sound absorption device is based on the fact that when an acoustic wave affects the panel surface, this surface vibrates in relation to what caused it, thereby originating a vibratory movement in which there is a transformation from acoustic energy into heat energy, resulting from different losses due to friction in the structure and in the air gap.

Acoustic panels or diaphrams are normally divided into two groups: plates and membranes depending on whether or not panel rigidity is taken into account. The theoretical analysis of panel behavior when affected by sound waves is complicated since this depends on many factors: loss factor, geometrical shapes and placement, sound incident angle, position of sound sources, etc. Therefore, the sound absorption coefficient of this type of absorbing device is practically determined by very simplified models which give the absorbent behavior in an approximate way. The usual hypotheses consider that the panel vibrates without being deformed and the space between the panels and the rigid surface behind is airtight. They do not consider the panel rigidity, its geometrical dimensions nor the way it is placed. The combination formed by the board and the air space behind acts as a mass-spring system. Given these simplifications, the resonance frequency of the device at which its sound absorption is maximum can be obtained by the empirical formula:

$$f_r = \frac{1}{2\pi}\sqrt{\frac{\rho c^2}{me}} = \frac{60}{\sqrt{me}} \quad \text{Hz} \tag{17}$$

where m is the mass of the panel per surface unit in kg/m^2, e is the depth of the air space in m, c is the velocity of sound propagation in air, and ρ represents the density of air in kg/m^3.

A simple equation that takes into account panel radiation is as follows:

$$f_r = \frac{60}{\sqrt{(\rho_m h + 0.6\sqrt{ab})e}} \quad \text{Hz} \tag{18}$$

ρ_m is the density of the panel in kg/m^3, h is its thickness in m, and a and b, are the length and width of the rectangular panel, respectively. The resonance frequency value obtained by the above formula is included among those that would be obtained by taking into account panel rigidity with simply supported edges and with clamped edges. Figure 12 shows the diagrams of an acoustic panel and its mechanical equivalent.

The use of acoustic panel devices is advantageous when it is necessary to increase the absorption of low frequencies. However, for frequencies higher than 500 Hz, it is more expensive than other absorbing materials or devices.

Acoustical panels have a small sound absorption coefficient. Moreover, if the panels are rigid, they are more selective.

The placing of absorbent material in the air space, either glued or not, increases the absorption coefficient of the device considerably and makes it less selective. Figure 13 shows the sound absorption coefficient of a plywood panel with a mass surface density of m = 2 kg/m^2, and different air-space thicknesses with and without porous material that is 25 mm thick.

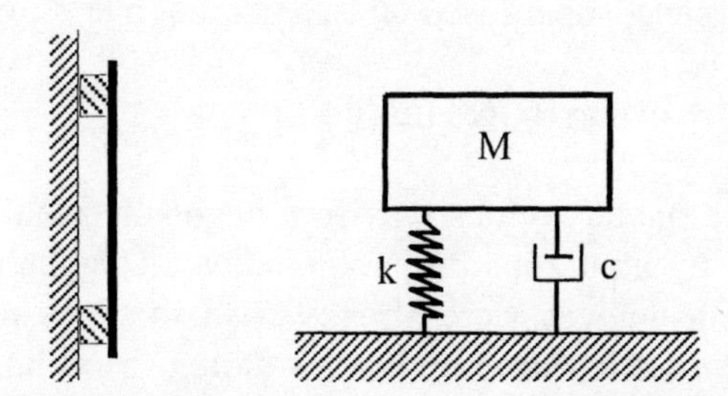

Figure 12. Mechanical equivalent of an acoustic panel resonator.

If the depth of the air space is very big, the sound absorption phenomenon of the panel is governed by its rigidity. In this case, for a supported rectangular panel with length *a*, width *b*, and thickness *h*, the fundamental resonance frequency can be expressed as follows:

$$f_r = \frac{\pi}{2}\sqrt{\left(\frac{1}{a^2}+\frac{1}{b^2}\right)\frac{Eh^2}{12(1-\nu)\rho_m}} \quad \text{Hz} \tag{19}$$

in which *E* is the Young modulus, ν is the Poisson coefficient and ρ_m is the density of the panel.

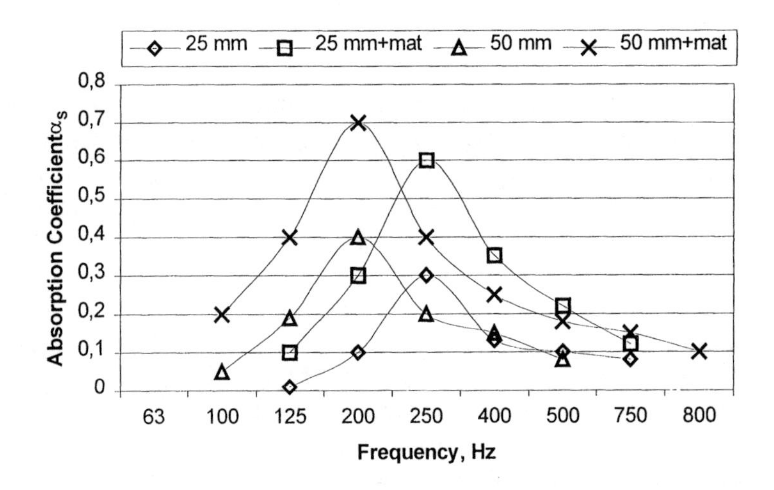

Figure 13. Sound absorption coefficient of a plywood board with and without absorbent material in the air space.

The panels can be combined in different ways in order to accommodate each particular case to the sound spectrum regions. For example, the panels can be juxtaposed on the same plane, superposed on two parallel planes, etc.

5.3 Helmholtz resonators or cavity resonators

The study of acoustic resonant cavities was first begun by Helmholtz and Rayleigh and has been continued by numerous research studies on the subject [34–36]. These types of sound absorbing devices were already used in times prior to Helmholtz´s work. They have been found in open air theatres dating from the classical periods of Greece and Rome and also in some medieval churches in northern Europe.

A Helmholtz acoustic resonator is composed of a cavity filled with air which communicates with the exterior through a neck. It absorbs sound energy around a specific resonance frequency, which is a function of the geometrical characteristics of the resonator. When sound waves reach the resonator, the air in the neck is put in motion, which in turn compresses and expands the air in the cavity. Due to friction on the walls of the neck, part of the sound energy dissipates in the form of heat. This loss of energy can be increased by introducing a very light porous material in the mouth of the neck, or an absorbent material in the cavity.

The Helmholtz principle of sound absorption by resonance can be applied basically to three types of devices or combinations of these: single resonators, perforated panels and alveolate resonators.

5.3.1 A single resonator in an infinite wall

From an acoustic point of view, Helmholtz resonators can be characterized by three magnitudes: cavity volume, V, length of the neck, l, and lateral area of the neck, S. Figure 14 shows a drawing of a Helmholtz resonator.

Helmholtz resonators are very selective. They reach their maximum sound absorption at the frequency of the resonance system. In the case of a single inlaid resonator with a cylindrical neck radius of a, this can be calculated with the following formula [37]:

$$f_{\mathrm{r}} = \frac{c}{2\pi}\sqrt{\frac{S}{V(l+1.6a)}} \quad \text{Hz} \tag{20}$$

in which c is the sound velocity in air. All the physical magnitude units are in accordance with the International System.

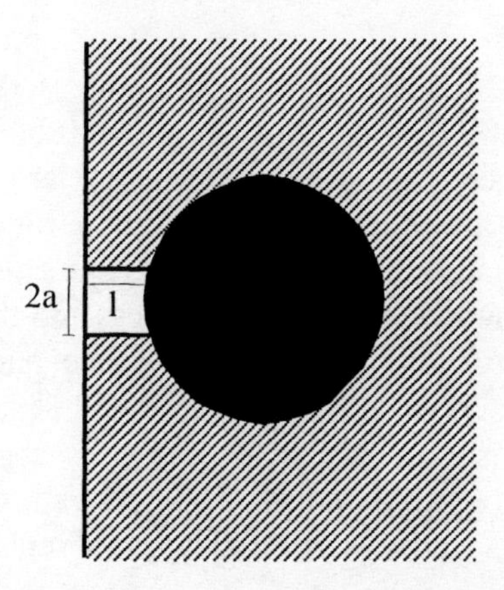

Figure 14. Helmholtz resonator.

The resonators can be designed to function at any frequency. It is a question of analyzing the appropriate dimensions. Generally, they are good absorbers at low frequencies. Their most normal use is to control the reverberation in an enclosure at low and medium frequencies.

It is theorically proven that maximum sound absorption of a single resonator is

$$A = \frac{c^2}{4\pi f_r^2} \quad \text{m}^2 \tag{21}$$

Figure 15 shows the sound absorbent characteristics of a single resonator with and without absorbent material in the cavity.

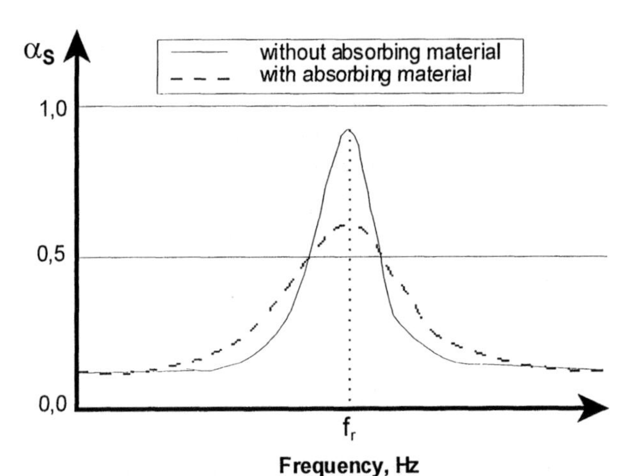

Figure 15. Sound absorbing characteristics of a single Helmholtz resonator.

5.3.2 Perforated panels

Single resonators are sound absorbing devices which are very selective in their range of frequencies. If there is a need to widen this range, the use of single resonators is not economically worthwhile. Therefore, multiple resonators are used. These consist of a perforated panel situated at a certain distance from a rigid wall. They act as an assembly of Helmholtz resonators in which each hole is equivalent to the neck of a resonator, the length corresponds to the panel thickness and the volume represents the quotient between the existing volume in the air space and the number of holes. Associated resonators are not as selective in their sound absorption. When absorbent material is placed in the air space, this factor increases absorption very effectively above and below resonance frequency values. The perforations are usually cylindrical or grooved. Figure 16 presents a cross-section diagram of a perforated panel.

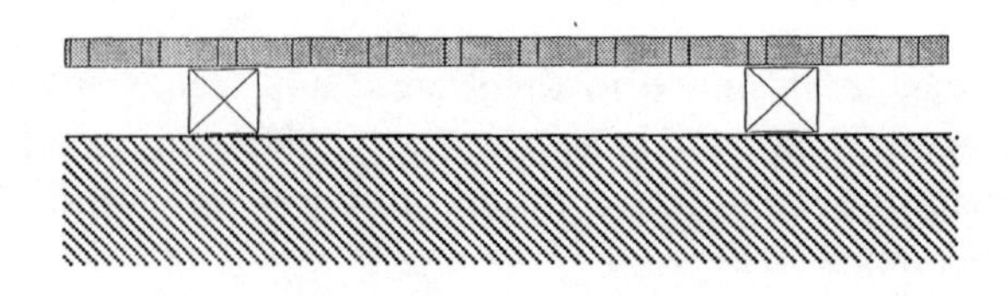

Figure 16. Perforated panel.

In the case of panels with cylindrical perforations situated at a distance, *d*, from a rigid wall, without sound absorbing material in the air space, the resonance frequency of the system can be calculated with the equation [38]

$$f_r = \frac{c}{2\pi}\sqrt{\frac{P}{d(l+1.6a)}} \quad \text{Hz} \tag{22}$$

in which *P* is the perforation coefficient, that is, the ratio between the sum of the perforation areas and the total area of the panel.

Figure 17 shows the sound absorption coefficient curves for different types of perforated panels with a 25 mm thick air space, filled with glass-wool.

The percentage of perforated area influences the sound absorption coefficient. Figure 18 shows the results obtained for a 2 mm thick perforated metal panel with a 75mm air space between the metal panel and the rigid wall, which is filled with mineral wool.

Another important factor is the thickness of the air space between the perforated panel and the rigid surface. The greater the thickness is, the more noticeable is the increase in sound absorption at low frequencies.

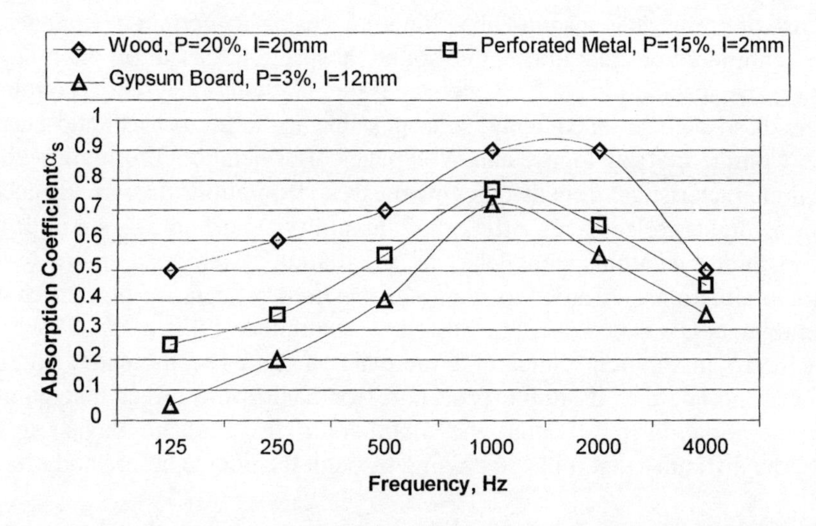

Figure 17. Sound absorption coefficients of several perforated panels.

5.4 Prefabricated sound absorbers

Manufactured absorbing devices are normally classified as tiles or boards, and special mountings. They are available in different sizes with a great variety of compositions, surface characteristics and mountings. These factors produce differences in the sound absorption coefficients, external appearance, flame resistance, light reflectance, cleaning and repainting possibilities, etc.

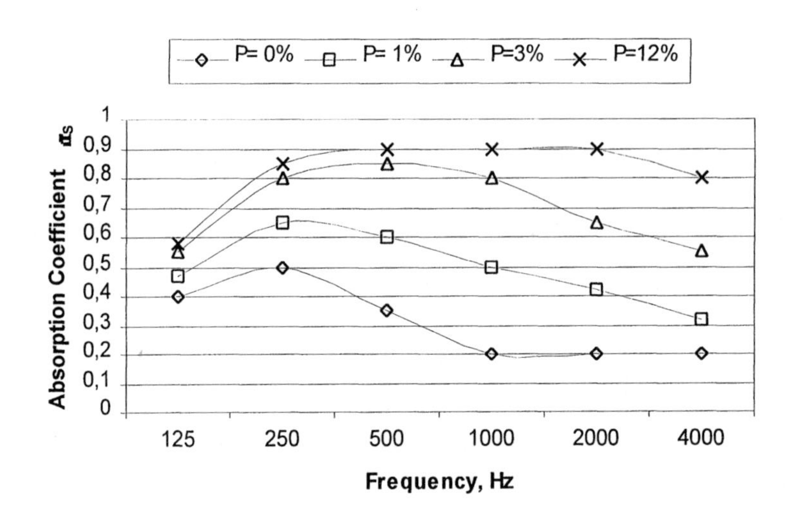

Figure 18. Effect of perforated area percentage on the sound absorption coefficient.

The use of prefabricated sound absorbing devices offers certain advantages. First of all, the manufacturer guarantees the sound absorption based on laboratory tests. Moreover, the installation and maintenance are relatively easy and economical. By choosing these devices adequately, it is possible to combine aesthetics, safety and comfort. Commercial catalogues and Web pages offer detailed information about their technical characteristics: dimensions, thicknesses, decoration, flame resistance, sound absorption, light reflectance, effects of humidity, mechanical strength, thermal insulation, microbial atmosphere behavior, installation system, etc.

In some situations, in order to maintain the light reflectance properties of sound absorbent devices, it is necessary to wash and repaint them. It is essential to follow the manufacturer´s instructions concerning the paints that are recommended and the types of application. There is a simple procedure for testing the effect that paint has on sound absorption: blow air from the mouth onto the acoustic device in order to compare the difficulty that it has in passing through the device before and after paint is applied.

The presence of an air space behind prefabricated sound absorbers significantly increases their sound absorption at low frequencies. Figure 19 illustrates this behavior in the case of suspended ceilings made of fiberglass acoustical tiles, 30 mm thick, with different finishes and different air space heights.

Acoustic tiles are manufactured in different thicknesses and densities. Figure 20 shows the sound absorption coefficient of three different thicknesses in the case of a suspended ceiling formed by rigid rockwool tiles which are covered with a thin coating of glasswool on both sides.

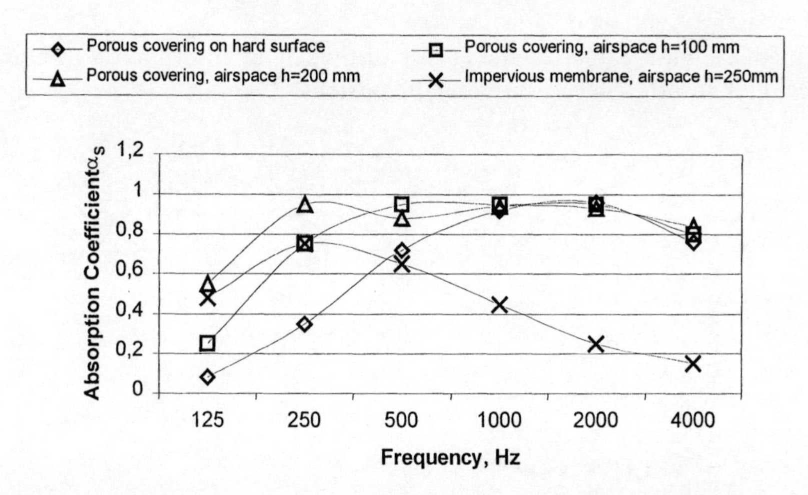

Figure 19. Influence of finishes and air space on a suspended ceiling.

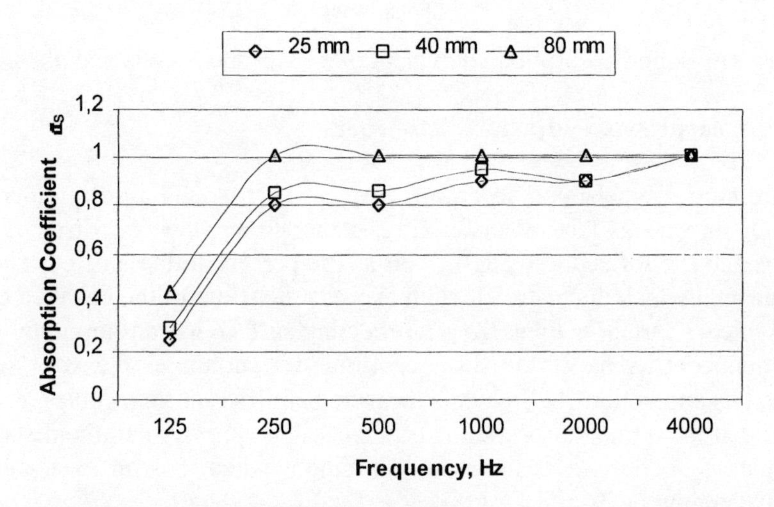

Figure 20. Influence of acoustic tile thickness on function of frequency.

5.5 Suspended acoustical absorbers

There are situations in which it is not possible to use conventional sound absorbing devices, or in which a great amount of sound absorption is needed in the enclosure. In these cases, functional, unit absorbers or suspended sound absorbing devices are used. These are hung from the enclosure ceiling.

They are made in different geometrical configurations and can provide significant sound absorption per unit. Although this varies according to the placement and proximity to other units. Manufacturers will advise the most efficient kind of placements. Figure 21 shows the curves of the sound absorption area equivalent of a vertical unit absorber in two different dimensions.

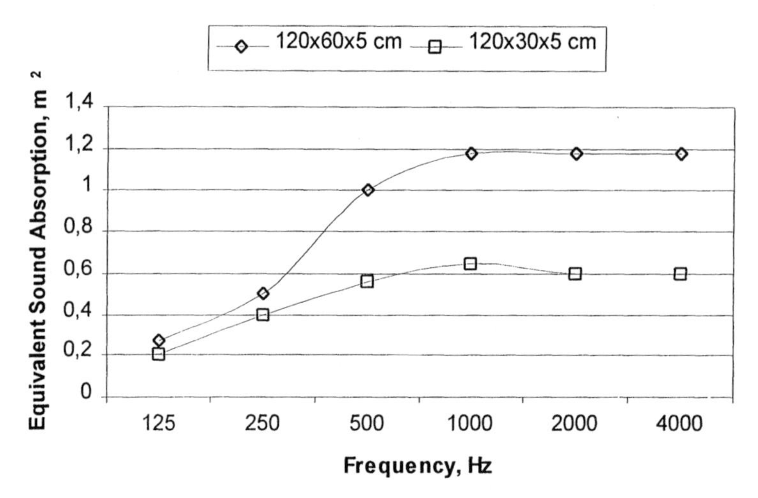

Figure 21. Sound absorption equivalent of a suspended acoustical absorber.

5.6 Curtains, carpets and adjustable absorbers

Curtains are porous-type sound absorbing devices. Their sound absorption coefficient [39] depends on various factors: material resistance to air flow, the distance between the curtains and the reflecting wall, as well as the percent fullness of curtain, that is, the percentage in which the material width exceeds the width of the installed curtain.

The distance a curtain is hung from a reflecting surface has an important effect on its absorption coefficient. If the distance from the curtain to the wall remains a constant and its absorption coefficient is measured at different frequencies, it will soon be obvious that maximum absorption coefficients will appear when the distance is an odd multiple of quarter-wavelengths and minimum when it is an even multiple of quarter-wavelengths.

Carpets are other normally used materials that offer sound absorption characteristics. Their sound absorption coefficients [40] depend on the height and weight of the pile, the type and thickness of the supporter and the floor on which they are lying.

Table 3 indicates some experimental results of sound absorption coefficient for curtains and carpets.

In the case of an enclosure with different uses, the optimum time reverberations and acoustic conditions are different for each one of them. Under these circumstances, it is possible to design different elements whereby variations in their position can increase or decrease the sound absorption in the enclosure. In reality, they require appropriate maintenance and, in many cases, have been replaced by electronic control of reverberation. Figure 22 shows some examples of adjustable absorbers: (a) retractable curtain, (b) hinged panels and (c) rotating elements.

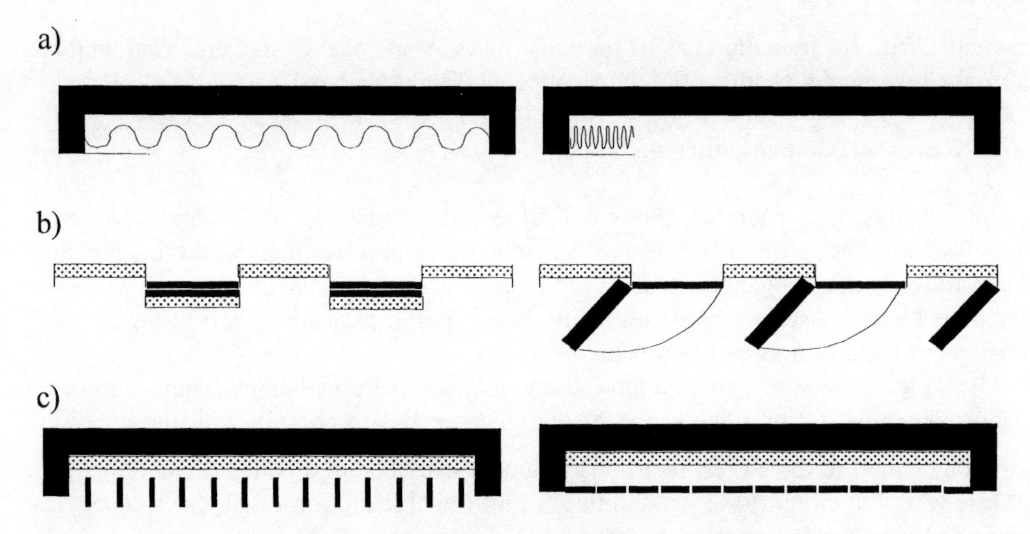

Figure 22. Examples of adjustable absorbers.

6 Active sound absorbers

Technological advances and, concretely, the appearance of fast digital signal processors, have turned into reality the ideas that were outlined by Lueg, in 1933, and ratified by Olsen, in 1953. These ideas are based on the assumption that it is possible to eliminate a noise by emitting, through a loudspeaker, another noise with similar characteristics, but in counterphase. This is what is known as active control of sound.

Active noise control systems are successfully employed to cancel noises at low frequencies, mainly in unidimensional sound fields, such as ducts. As for tridimensionals, active control applications are centered on the elimination of discrete modes at low frequency, and in the production of localized areas of silence.

There have also been attempts to apply active technologies to sound absorption control in an enclosure. In this respect, satisfactory results have been obtained in very specific test conditions. However, for the moment, they are impractical for the real conditions of an enclosure [41, 42].

7 Measurement of sound absorption coefficients

The sound absorption coefficients of sound absorbent materials or devices are difficult to determine in the actual mounted conditions. Nevertheless, it is important to know their approximate values, even if they are obtained in laboratory conditions. Based on this information, it is then possible to design the control or correction of some acoustic parameters in the enclosures.

There are several methods for determining sound absorption coefficients. However, at present, there is no procedure that relates the measurement results obtained for each one of them. Each method has its particular uses. Some of the most common methods are the reverberant chamber, the impedance tube and tone burst.

7.1 Reverberation chamber method

The reverberation chamber method [25] entails measuring the sound absorption coefficient of acoustic materials used as wall and ceiling treatments, or the equivalent absorption area of objects such as furniture, people or functional absorbers. The results obtained can be used for comparative or design purposes, with regard to the acoustic behavior of enclosures and noise control.

In order to carry out this method, the volume of the reverberant chamber must be approximately 200 m^3. The shape of the chamber has to comply with the condition that the length of the largest rectilinear segment inscribed in it be less than $1.9\sqrt[3]{V}$ m. Moreover, for the purpose of attaining a uniform distribution of modal frequencies, especially in the low frequency bands, the dimensions of the sides must not be the same or be in the proportions of the first integer numbers. The sound field has to be as diffuse as possible and the equivalent absorption area of the empty chamber must be very small.

The sample, if plane, must have an area between 10 and 12 m^2, and be mounted according to specifications. The tests with separate absorbers (chairs, people, special absorbers, etc.) need to comprise a sufficient number of objects distributed randomly throughout the enclosure to produce a measurable change in the equivalent absorption area.

The measurements must be performed in third-octave bands with central frequencies ranging from 100 to 5000 Hz. The sound absorption coefficient α_S of a plane absorber should be calculated with the formula

$$\alpha_S = \frac{55.3V}{cS}\left(\frac{1}{T_2} - \frac{1}{T_1}\right) \tag{23}$$

in which V is the volume of the chamber in m^3; c is the velocity of sound in the air in m/s; S is the area of the test sample in m^2; T_1 and T_2 are the reverberation times, in seconds, of the reverberant chamber, when empty, and with the test sample, in that order.

The chamber method enables the sound absorption coefficient for natural size samples to be calculated in situations similar to reality. It uses the Sabine reverberation time formula, which is obtained from hypotheses that are not always fulfilled in the enclosure. All sound absorption coefficients supplied by manufacturers for use in architectural acoustic calculations are measured by the reverberation chamber method. It is common practice to list the coefficients in six octave bands with central frequencies ranging from 125 to 4000 Hz.

7.2 Impedance tube method

The impedance tube method, or Kundt tube, permits small samples to be measured for the following: the sound absorption coefficient of a material for normal incidence on its surface, the acoustic reflection coefficient, and normalized impedance.

The sample to be analyzed is put in a tube which is closed at one end. At the other end, a loudspeaker emits sound waves. As a result, stationary waves are produced in the tube from the composition of the incident and reflected plane waves. These waves are characterized by having fixed maxima and minima in space, and are separated by a quarter of a wavelength. In the maxima, incident and reflected waves are in phase, whereas in the minima they are in opposite phase.

The reflection coefficient and acoustic impedance can be evaluated by measuring the pressure in the tube in different positions [43, 44]. At frequencies lower than 200 Hz, the reflection coefficient of porous samples in the Kundt tube is very close to 1. The evaluation of acoustic impedance is affected by systematic errors which are difficult to cancel. There is, however, a method with two microphones and three calibrations [45] that improves the precision of these measurements significantly.

Some of the limitations of this method include: the sound absorption coefficient that is measured is at normal incidence; only small samples can be tested; and samples of vibrating plates cannot be measured, since their resonance frequencies depend on the dimensions and shape of their supporter. Figure 23 shows a cross-section diagram of the impedance measurement tube.

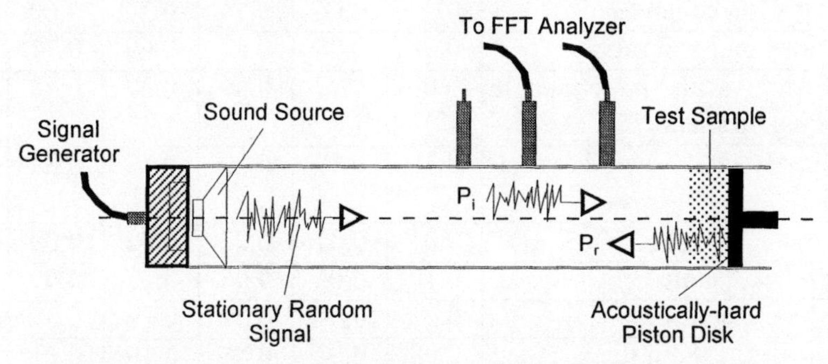

Figure 23. Cross-section diagram of the impedance tube.

7.3 Tone-burst method

The tone-burst method can be used to measure the sound absorption coefficient of a material at any desired angle of incidence. The method consists of emitting a desired frequency tone into an enclosure for a very short time interval, which is then received by a microphone at a known distance and, finally, the sound pressure level is measured. Afterwards, the loudspeaker is aimed at the test specimen, forming a desired angle of incidence, whereby the total path length of the reflected sound is the same as the one between the source and the microphone. Then, the tone burst is once again emitted. By comparing the two sound pressure levels obtained from the direct and reflected sound for the same path length, the reflection coefficient can be determined for a particular angle of incidence and at a specific frequency. Thereafter, the sound absorption coefficient can be calculated.

This method can be applied "in situ" but it has important limitations at frequencies lower than 1000 Hz.

8 Sabine sound absorption coefficients for some common building materials and furnishings

Table 3 lists Sabine sound absorption coefficients for some common building materials and furnishings, which are obtained from several references [39, 40, 46, 47]. It does not include general commercial acoustical materials, since manufacturers provide their own sound absorption coefficients, mounting specifications, and other properties. The values shown are approximate.

Table 3. Sound absorption coefficients for some common building and furnishings.

Material	Frequency center octave band, Hz					
	125	**250**	**500**	**1000**	**2000**	**4000**
Brick, unglazed	0.02	0.02	0.03	0.04	0.05	0.07
Brick, unglazed and painted	0.01	0.01	0.02	0.02	0.02	0.03
Carpet heavy, on concrete	0.02	0.06	0.14	0.37	0.60	0.65
Carpet heavy, on foam rubber	0.08	0.24	0.57	0.69	0.71	0.73
Carpet heavy, with impermeable latex backing on foam	0.08	0.27	0.39	0.34	0.48	0.63
Chair, metal or wood seat, unoccupied	0.15	0.19	0.22	0.39	0.38	0.30
Concrete	0.01	0.01	0.02	0.02	0.02	0.02
Concrete block, painted	0.10	0.05	0.06	0.07	0.09	0.08
Concrete block, coarse	0.36	0.44	0.31	0.29	0.39	0.25
Cork floor tiles, 1.9 cm thick, glued down	0.08	0.02	0.08	0.19	0.21	0.22
Curtains, lightweight, 338 g/m^2, flat on wall	0.03	0.04	0.11	0.17	0.24	0.35
Curtains, mediumweight, 475 g/m^2, 100% fullness	0.07	0.31	0.49	0.75	0.70	0.60
Curtains, heavyweight, 610 g/m^2, 100% fullness	0.14	0.35	0.55	0.72	0.70	0.65
Fiberglass or rockwool boards and blankets, 2.5 cm, 24–48 kg/m^3, on rigid solid surface	0.01	0.25	0.65	0.85	0.80	0.75

Table 3 (continued)

Material	Frequency center octave band, Hz					
	125	**250**	**500**	**1000**	**2000**	**4000**
Fiberglass or rockwool boards and blankets, 5 cm, 24–48 kg/m^3, on rigid solid surface	0.17	0.55	0.80	0.90	0.85	0.80
Glass: heavy, large panes	0.18	0.06	0.04	0.03	0.02	0.02
Glass: ordinary window	0.35	0.25	0.18	0.12	0.07	0.04
Grass, 5 cm high	0.11	0.26	0.60	0.69	0.92	0.99
Gravel, loose and moist,10 cm thick	0.25	0.60	0.65	0.70	0.75	0.80
Gypsum board, 1.27 cm nailed to 5.1 by 10.2 cm studs 41 cm center to center	0.29	0.10	0.05	0.04	0.07	0.09
Linoleum, rubber, asphalt or cork tile on concrete	0.02	0.03	0.03	0.03	0.03	0.02
Marble or glazed tile	0.01	0.01	0.01	0.01	0.02	0.02
Mineral fiber, 1.27 cm thick, spray-on materials	0.05	0.15	0.45	0.70	0.80	0.80
Mineral fiber, 2.5 cm thick, spray-on materials	0.16	0.45	0.70	0.90	0.90	0.85
Plaster, gypsum or lime, smooth finish on tile or brick	0.013	0.015	0.02	0.03	0.04	0.05
Plywood paneling, 1 cm thick	0.28	0.22	0.17	0.09	0.10	0.11
Polyurethane foam, 27 kg/m^3, 1.5 cm thick	0.08	0.22	0.55	0.70	0.85	0.75
Rough soil	0.15	0.25	0.40	0.55	0.60	0.60
Steel	0.05	0.10	0.10	0.10	0.07	0.02
Terrazzo	0.01	0.01	0.02	0.02	0.02	0.02
Water surface (swimming pool)	0.01	0.01	0.01	0.02	0.02	0.03
Wood, floor	0.15	0.11	0.10	0.07	0.06	0.07
Wood parquet on concrete	0.04	0.04	0.07	0.06	0.06	0.07
Wood panel, 1 to 1.3 cm thick, over 5 to 10 cm air space	0.30	0.25	0.20	0.17	0.15	0.10
Wood platform with large space below	0.40	0.30	0.20	0.17	0.15	0.10

References

[1] Morse, P.M. and Ingard, K.U., *Theoretical Acoustics,* McGraw-Hill. Pbk Reprint Princeton University Press, p. 580, 1986.

[2] ISO. Attenuation of sound during propagation outdoors, Part 1, ISO/DIS 9613-1. International Organization for Standardization, CH-1211 Geneva 20, Switzerland, 1990.

[3] Beranek, L.L., Audience and chair absorption in large halls:II, *J.Acoust. Soc. Am.*, **45**, pp. 13-19, 1969.

[4] Kosten, C.W., New method for the calculation of the reverberation time of halls for public assembly, *Acústica*, **16**, pp. 325-330, 1965.

[5] Beranek, L.L., *Concert and Opera Halls: How They Sound.* Published for the Acoustical Society of America through the American Institute of Physics, Woodbury NY, pp. 621-627, 1996.

[6] Sabine, W.C., *Collected Papers on Acoustics*, Dover, New York, 1964.

[7] Franklin, W.S., Derivation of equation of decaying sound in a room and definition of open window equivalent of absorbing power, *Phys. Rev.*, **16**, pp. 372-374, 1903.

[8] Jaeger, G., *Toward a Theory of Reverberation*, Sitzungsber. Kais. Akad. Wiss. Vienna. Math.-Naturw, Klasse, Bd. 120 Abt. IIa, pp. 613-634, 1911.

[9] Buckingham, E., Bur. Standards, Sci. Paper, n° 506, 1925.

[10] Eyring, C.F., Reverberation time in “dead” rooms, *J. Acoust. Soc. Am.*, **1**, pp. 217-241, 1930.

[11] Knudsen, V.O., Appendix II. *Architectural Acoustics,* Wiley, New York, pp. 603-605, 1932.

[12] Schuster and Waetzmann, *Ann. d. Phys*., March, 1929.

[13] Milligton, G., A modified formula for reverberation, *J. Acoust. Soc. Am.*, **9**, 1932.

[14] Jordan, V.L., Room acoustics and architectural development in recent years, *Appl. Acoust*, **2**, pp. 59-81, 1969.

[15] Davis, D. and Davis, C., *Sound System Engineering*, Howard W. Sons and Company, pp. 191-201, 1997.

[16] Lazarus, H., Prediction of verbal communication in noise - a development of generalized SIL curves and the quality of communication, *Appl. Acoust.*, **20**, pp. 245-261, 1987.

[17] Kryter, K.D., Methods for the calculation and use of the articulation index, *J. Acoust. Soc. Am.*, **34**, pp. 1689-1697, 1962.

[18] Peutz, U.M.A., Articulation loss of consonants as a criterion for speech transmission in a room, *J. Audio Eng. Soc.*, **19**, pp. 426-454, 1971.

[19] IEC Std.268-16, The Objective Rating of Speech Intelligibility in Auditoria by the RASTI- Method, 1987.

[20] Reichardt, W., Alim, D.A., and Schmidt, W., Definition und Messgrundlungen eines objektiven Masses zur Ermittlung der Grenze zwischen brauchbarer und unbrauchbarer Durchsichtigkeit bei Musikdarbeitungen, *Acustica*, **32**, pp. 126-137, 1975.

[21] Thiele, R., Richtungsverteilung und Zeitfolge der Schallrückwürfe in Räumen, *Acustica*, **3**, pp. 100-112, 1953.

[22] Kürer, R., Zur Gewinnung von Einzahlkriterien bei Impulsmessung in der Raumakustik, *Acustica*, **21**, pp. 370-372, 1969.

[23] Jordan V., *Acoustical Design of Concert Halls and Theatres,* Applied Science Publishers, London, 1980.

[24] Cremer, L. and Muller, H.A., *Principles and Applications of Room Acoustics*, Applied Science Publishers, London, 1982.

[25] Measurement of Sound Absorption in a Reverberation Room, ISO 354:1985, International Organization for Standardization, CH-1211 Geneva 20, Switzerland.

[26] Sound Absorption and Sound Absorption Coefficients by the Reverberation Room Method, ASTM C423, American Society for Testing and Materials, Philadelphia, PA 19103, USA.

[27] Mounting Test Specimens During Sound Absorptive Tests, ASTM E795, American Society for Testing and Materials, Philadelphia, PA 19103, USA.

[28] Zwikker, C. and Kosten, C.W., *Sound Absorbing Materials,* Elsevier, New York, 1949.

[29] Materials for Acoustical Applications. Determination of airflow resistence. ISO 9053:1991. International Organization for Standardization, CH-1211, Geneva 20, Switzerland.

[30] Delany M.E. and Bazley, E.N., Acoustical properties of fibrous absorbent materials, *Appl. Acoust.*, **3**, pp. 105-116, 1970.

[31] Bies, D.A. and Hansen, C.H., Flow resistance information for acoustical design, *Appl. Acoust.*, **13**, pp. 357-391, 1980.

[32] Attenborough, K., Acoustical characteristics of porous materials, *Phys. Rep.*, **82** pp. 179-227, 1982.

[33] Attenborough, K., On the acoustic slow wave in air-filled granular media, *J. Acoust. Soc. Am.*, **81**, pp. 93-102, 1987.

[34] Pedersen, P.O., Lydtekniske Undersøgelser (Acoustic investigations), Scient. Eng. Papers, 1940-1945.

[35] Ingard, U., On the theory and design of acoustic resonators, *J. Acoust. Soc. Am.*, **25**, pp. 1037-1061, 1953.

[36] Ingard, U., Perforated facing and sound absorption, *J. Acoust. Soc. Am.*, **26**, pp. 151-154, 1954.

[37] Beranek, L.L. and Vér, I.L., *Noise and Vibration Control Engineering. Principles and Applications,* John Wiley & Sons, New York, pp. 234-235, 1992.

[38] Bruel, P.V., *Sound Insulation and Room Acoustics,* Chapman & Hall Ltd, p. 118, 1951.

[39] Mankovsky, V.S., *Acoustics of Studios and Auditoria*, Focal Press Ltd, London, 1971.

[40] Carpet Specifier´s Handbook, Chap.9, American Carpet and Rug Institute, Dalton, GA30722, USA.

[41] Orduña-Bustamante, F. and Nelson, P.A., An adaptative controller for the active absorption of sound, *J. Acoust. Soc. Am.*, **91**, pp. 2740-2747, 1992.

[42] Ruppel, T. and Shields, F.D., Cancellation of airborne acoustic plane waves obliquely incident upon planar phased array of active surface elements, *J. Acoust. Soc. Am.*, **93**, pp. 1970-1977, 1993.

[43] American Society for Testing and Materials, Test Method for Impedance and Absorption of Acoustics Materials by the Tube Method. ASTM C384-77, 1977.

[44] ASTM Committee E-33 on Environmental Acoustics, American Society for Testing and Materials, Philadelphia, PA. Impedance and absorption of acoustical materials using a tube, two microphones, and a digital frequency analysis system.

[45] Gibiat, V. and Laloë, F., Acoustical impedance measurements by the two-microphone-three calibration (TMTC) method, *J. Acoust. Soc. Am.*, **88**, pp. 2533-2545, 1990.

[46] Harris, C.M., *Noise Control in Buildings. A Practical Guide for Architects and Engineers,* McGraw-Hill, Inc., New York, 1994.

[47] Hedeen, R.A., *Compendium of Materials for Noise Control,* National Institute for Occupational Safety and Health, Publication 80-116, Cincinnati, USA, May, 1980.

Chapter 3

3-D Sound and auralization

D.R. Begault
San Jose State University, Human Factors Research and Technology Division, NASA Ames Research Center, Moffett Field, CA 94035-1000, USA
Email: db@eos.arc.nasa.gov

Abstract

An introduction to the basics of auditory localization and 3-D sound is overviewed for application to auralization, "the process of rendering audible, by physical or mathematical modeling, the sound field of a source in a space, in such a way as to simulate the binaural listening experience at a given position in a modeled space." The essential components of the modeling are reviewed and perceptual limitations of the simulation are outlined.

1 Introduction

"3-D sound" is a generic term that describes a host of digital computational techniques for rendering a virtual acoustic source at a desired location. The technology has only recently made the transition from the laboratory to the commercial audio world, and has since found application primarily to gaming and entertainment applications. However, there are several less-publicized application areas in the domain of human–computer interaction, and specifically to virtual acoustic rendering of an acoustic space, or *auralization.* Auralization has been defined as "the process of rendering audible, by physical or mathematical modeling, the sound field of a source in a space, in such a way as to simulate the binaural listening experience at a given position in a modeled space" (Kleiner, *et al.* [1]). Prior to describing this auralization, it is necessary to describe the basics of auditory localization and its implementation in 3-D audio systems (see also Begault [2]).

2 Basics of localization

Nominally, a sound is described in terms of three perceptual descriptors: pitch, tone color and loudness; or in terms of their equivalent physical descriptors, frequency, spectral content and intensity. However, spatial location is also an important perceptual attribute. For a long time, researchers held that inter-aural level and inter-aural time differences were the primary cues used for auditory localization. For instance, if you snap your fingers to the right of your head, the wavefront's amplitude will be greater at the right ear than at the left ear. Higher frequencies are shadowed from the opposite ear by the head; but frequencies below about 1.5 kHz diffract around to the opposite ear. The difference in levels at the two ears is interpreted as changes in the sound source position from the perspective of the listener. This is the same cue used by the balance knob on a stereo playback system; by adjusting the output level between left and right speakers, it is possible to manipulate the perceived spatial location of the virtual image. Note that when the signal is equal from both speakers and one listens from a position between them, the signal appears at a virtual position between the speakers, rather than as two sounds in either speaker.

Another cue for spatial hearing is the inter-aural time difference. The wavefront of the finger snap reaches the right ear before the left ear in time, since the path length to that ear is shorter, and sound travels at a constant rate through air. These differential arrival times are also evaluated as a cue to the change of a sound source's position. Wearing headphones and using a digital delay applied to the left channel, one can make a recording of a finger snap travel from the center of the head to the right side by increasing the inter-channel delay from 0 to approximately 1 ms. This time delay cue is most effective for lower frequencies below 1.5 kHz, but also functions to some degree across the entire audible range.

Effects produced like this over headphones are not really localized as much as they are lateralized. This refers to spatial illusions that are heard inside or at the edge of the head, but never seem at a distant point outside the head, or externalized. This is for three reasons. One is the lack of proper feedback from head motion cues. When localizing sound, we move our heads in order to minimize the inter-aural differences, i.e. using the head as a sort of acoustic pointer. This makes sense because we usually use our aural and visual senses together, and want to look at the object localized via sound (probably to decide whether to pounce or run, speaking from an evolutionary perspective). With headphone listening, head movement causes no change in spatial auditory perspective – it remains invariant, no matter where your head is. Head movement in relationship to stereo loudspeakers outside the head is even worse because intensity and time differences change with no relationship to the intended virtual imagery.

Another reason for lateralization with only intensity and time differences has to do with the lack of the spectral modification caused by the outer ears (the pinnae). One can think of this spectral modification as equivalent to what a graphic equalizer does– emphasizing some frequency regions while attenuating others. The modification at one ear for a given position is technically referred to as the head related transfer function

(HRTF). This spectral modification was only recently recognized as an important cue to spatial hearing, especially for front–back and up–down discrimination.

Consider a sound source at right 60 degrees azimuth, 0 degrees elevation; and another at the same elevation but at a "mirror image" position of right 120 degrees azimuth. These two sound sources have roughly the same overall inter-aural time and intensity differences, as shown in Figure 1. However, the rearward sound will have a relatively "duller" timbre. In fact, for a given broad band sound source, each elevation and azimuth position relative to the listener contains a unique set of HRTF-based spectral modifications that act as an acoustic "thumbprint," as shown in Figure 2. This is explained by the complex construction of the outer ears, which impose a set of minute delays that collectively translate into a particular two-ear (binaural) HRTF for each sound source position. In summary, there are three important cues for synthesizing a sound source to a given virtual position outside a headphone listener: (i) overall inter-aural level differences; (ii) overall inter-aural time differences; and (iii) spectral changes caused by the HRTF. Finally, it is important to recognize that spatial hearing occurs within an overall context of other perceptual cues, primarily from vision, and that memory and expectation can modify acoustic cues.

Spatial HRTF effects can be captured for a fixed listening position by making a binaural (dummy head) recording of a sound source, and then listening later to the recording through headphones. These mannequin heads contain a stereo microphone pair located in a position equivalent to the entrance of the human ear canal. One can obtain a realistically spatial recording with such a device, but it is then difficult to manipulate a particular virtual image to an arbitrary spatial position during post-production.

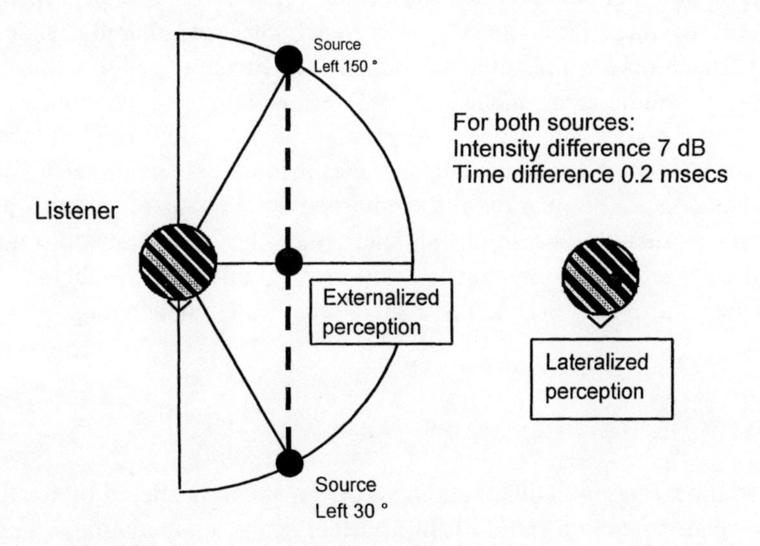

Figure 1. Sound source positions can have ambiguous percepts that are either externalized or lateralized within the head.

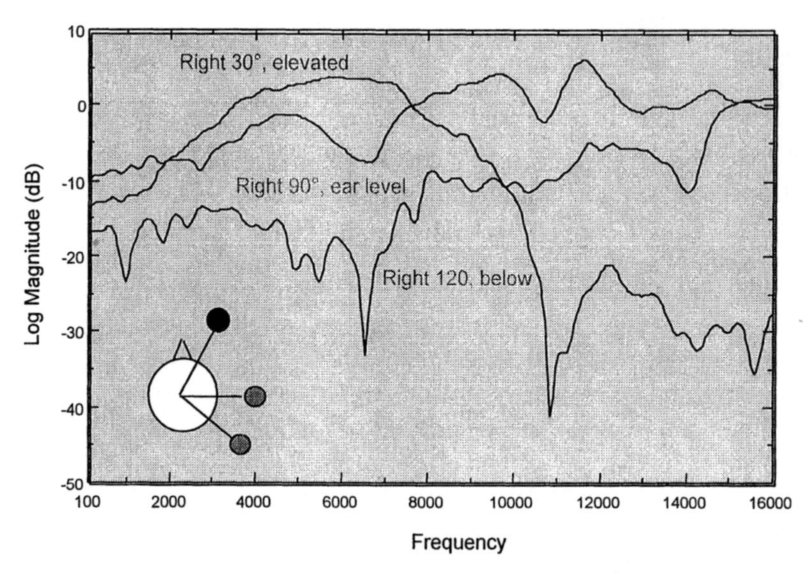

Figure 2. HRTF frequency modification for three positions. Inset: overhead view.

This limitation is overcome through the use of 3-D sound digital signal processing; either using specialized hardware or the "native" processor. Without going into detail, it is sufficient to know that one can take measurements of HRTF at various positions, and then translate these into digital filter parameters. These measurements can be made on a real or dummy head or on the specific ears of the listener for optimal fidelity. These parameters can then be selectively recalled by the signal processing apparatus. By a time-domain process known as convolution, a prerecorded or live sound is filtered by the HRTF filters (one for each ear), and thereby rendered to a specific spatial position for a headphone listener (see Begault [2] for further details). The 3-D sound system performs the necessary filtering and positional interfacing.

Figure 3 summarizes the aspects of spatial hearing that can be potentially manipulated by a 3-D audio system. This includes azimuth, elevation, and distance of the virtual sound source. Simulation of the environmental context is another aspect of virtual acoustic simulation, involving simulation of the virtual sound source size (auditory source width) and envelopment. In reality, all of these factors interact, making absolute control over 3-D audio imagery technically challenging.

3 Auralization

In natural spatial hearing, both direct and reverberant sound is altered by the listener's head related transfer function (HRTF). In a virtual acoustic simulation, when HRTF filtering is applied not only to the direct sound but also to the indirect sound, the result is spatial reverberation. The binaural room impulse response from a dummy head recording can capture this information, but only within a pre-existing room at a predetermined measurement point with a given set of outer ears. It is more desirable to obtain a synthetic impulse response of a real or simulated environmental context so that this information can be processed binaurally using 3-D sound system for real-time simulation.

Auralization involves the combination of room modeling programs and 3-D sound-processing methods to simulate the reverberant characteristics of a real or modeled room acoustically. The development of auralization systems has primarily been intended for the acoustical consultant to complement acoustic modeling software. Although acoustical parameters can be predicted using only software, auralization allows the consultant or their clients to listen to the effect of an acoustical treatment to a building. This application has had a long history (see, e.g. Horrall [3]; Kuttruff [4]). Another application is the creation of a virtual acoustic sound field that can be tied in with virtual reality-like walk throughs of acoustical spaces (e.g. Takala *et al.* [5]).

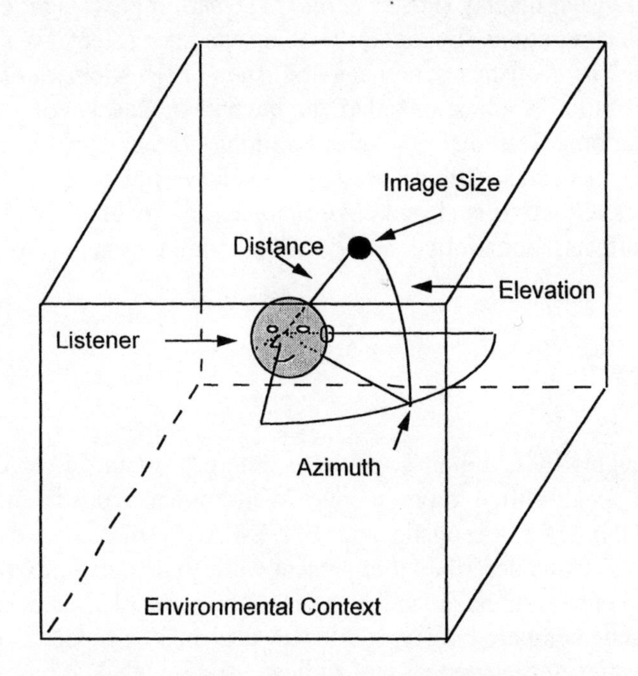

Figure 3. A taxonomy of spatial hearing.

Acoustical consulting applications can tolerate a certain level of difference between measured and synthesized signals. What matters is perceptual, not acoustical, equivalence between the signals. Nevertheless, the problem common to all of these applications is that the multiple source generation necessary for simulating a diffuse sound field can easily overwhelm computational resources. For instance, a simple image model of a rectangular room with early reflections up to fifth order results in over 200 valid sources (Jan and Flanagan [6]) (with very unrealistic sounding reverberation!). Using the computational savings of overlap-add fast fourier transform (FFT) techniques, two channels of a 3 s impulse response still require around 22 million floating point operations (Lehnert and Blauert [7]).

Fundamentally, all acoustic modeling programs and consequently any resulting auralizations will be dependent on the accuracy of several modeling parameters that are often difficult to characterize accurately. These include sound source directivity, absorption coefficients and diffusion. Modeling of coupled reverberant spaces can also be problematic. The results of an auralization system may not produce anything that sounds like the actual room being modeled because of the difficulty of merging accurate acoustical and perceptual models. The psychoacoustics of many of the issues surrounding room modeling and auralization–in particular the compromises necessary for realizable systems–are just beginning to be evaluated (e.g. Begault [8]; Jørgensen *et al.* [9]).

There are other uses for auralization that currently are receiving less attention than acoustical consulting, but are nonetheless important. This list includes advanced reverberation for professional audio applications (Kendall and Martens [10]); interactive virtual acoustic environments, based on information from a virtual world seen through a visual display (Foster *et al.* [11]); and improvement of localization for virtual acoustic simulations (Begault [12]). One can even select a seat in a concert hall by "previewing" its acoustics (Korenaga and Ando [13]). Kleiner *et al.* [1] have given a list of auralization applications that go beyond the near-term domain of sound system evaluation. The list includes training (architects, acousticians, audio professionals, musicians, blind persons); factory noise prediction; studies in psychoacoustics related to enclosed environments; microphone placement evaluation; video game effects; automotive and in-flight sound system design; and cockpit applications.

4 Overview

Auralization begins by using an acoustic modeling program to represent a particular space through a simplified room model. Well-known acoustic modeling progams include ODEON, CATT-Acoustic and EASE-EARS. In the software, a series of planar surfaces must be described to represent walls, doors, ceilings, and other features of the environmental context. Usually, the modeling must be restricted in the number of surfaces used, compared to typical CAD drawings. Second, the absorptive and diffusive properties of the surfaces and audience are modeled, at best usually in octave bands from 0.1 to 4 kHz. Calculation of the frequency-dependent magnitude and phase transfer function of a surface made of a given material will vary according to the size of the surface and the angle of incidence of the waveform (see Ando [14]; D'Antonio *et al.* [15]).

Once the software has been used to specify the details of a modeled room, sound sources may be placed in the model. Most sound system design programs are marketed by loudspeaker manufacturers, whose particular implementations are oriented around sound reinforcement; therefore, the level of detail available for modeling speakers, including aiming and dispersion information, is relatively high. After the room and speaker parameters have been joined within a modeled environment, details about the listeners can then be indicated, such as their number and position.

Prepared with a completed source–environmental context–listener model, a synthetic room impulse response can be generated by the room modeling software. A specific timing, amplitude and spatial location for the direct sound and early reflections is obtained, based on the ray-tracing or image model techniques described below. Usually, the early reflection response is calculated only up to around 100 ms. The calculation of the late reverberation field usually requires some form of approximation due to computational complexity. A room modeling program becomes an auralization program when the room impulse response is spatialized by a real-time convolution system, such as the Lake HURON. Auralization is also possible by performing a non-real-time convolution with a .wav or similar sound file format and then listening using standard playback software.

Figure 4 shows the basic process involved in a computer-based auralization system intended for headphone audition. The model also will include details about relative orientation and dispersion characteristics of the sound source, information on transfer functions of the room's surfaces, and data on the listener's location orientation, and HRTFs.

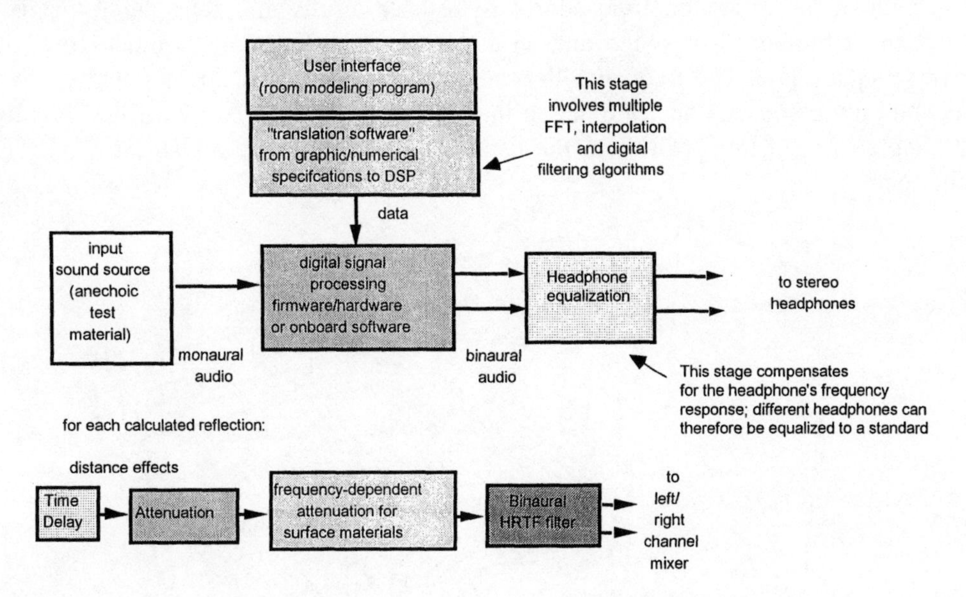

Figure 4. The basic components of the signal processing used in an auralization system.

Figure 5 shows a perspective view of a concert hall produced by a room modeling program capable of auralization (CATT-Acoustic; see http://www.netg.se/~catt). The dashed lines intersect on the location of a modeled sound source, while the numbers refer to five seating locations to be analyzed. Each surface is modeled according to its particular frequency-dependent absorption and diffusion characteristics. Figure 6 shows how the software allows the possibility of viewing an enclosure from different perspectives, via a mouse interface (this example shows the interior of a church). One can navigate through and around an enclosure, in a manner similar to exploration within a virtual world. In fact, VRML (Virtual Reality Modeling Language) output is also possible.

Once all of the parameters are determined for an enclosure such as shown in Figures 5 and 6, it is possible to execute an acoustical analysis of the early reflections and dense reverberation pattern for the indicated seating positions. The inclusion of HRTF information for the early reflections allows the derivation of a binaural impulse response for virtual acoustic modeling.

Figure 7 shows a section view of the hall that can be thought of as a "spatial calculation" of the early reflections, as determined by image model ray tracing. The size of the circle corresponds to the reflection intensity; the distance corresponds to time delay; and the location corresponds to the relative angle of incidence to the listener. Figure 8 shows the resulting binaural impulse response; the large peaks correspond to strong reflections within the enclosure.

At the bottom of Figure 4 is an example of the calculations used in determining the scaling, time delay and filtering of the direct sound and each modeled early reflection. For each reflection, a “copy” of the direct sound is obtained from a tapped delay line. The path of the reflection from source to listener results in a time delay that is a function of the speed of sound and an attenuation coefficient determined from the inverse square law. The frequency-dependent absorption of the room’s surfaces will depend on a complex transfer function that can be approximated by a filter. Finally, the angle of reflection relative to the listener’s orientation is simulated by a HRTF filter pair.

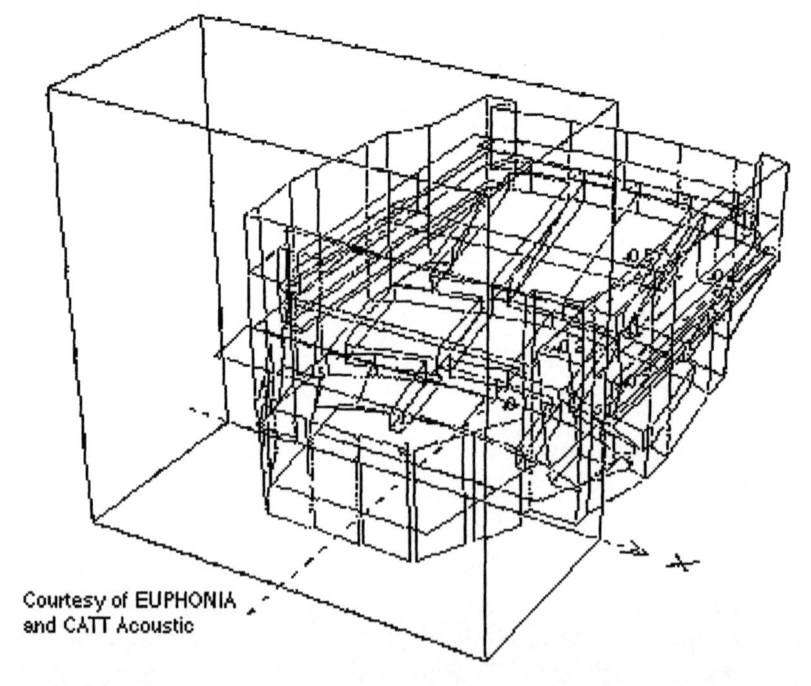

Figure 5. Perspective view of a hall: the Espace des Arts (Chalon-sur-Saôn, France: A. Moatti, architect and theater consultant; B. Suner, acoustical consultant).

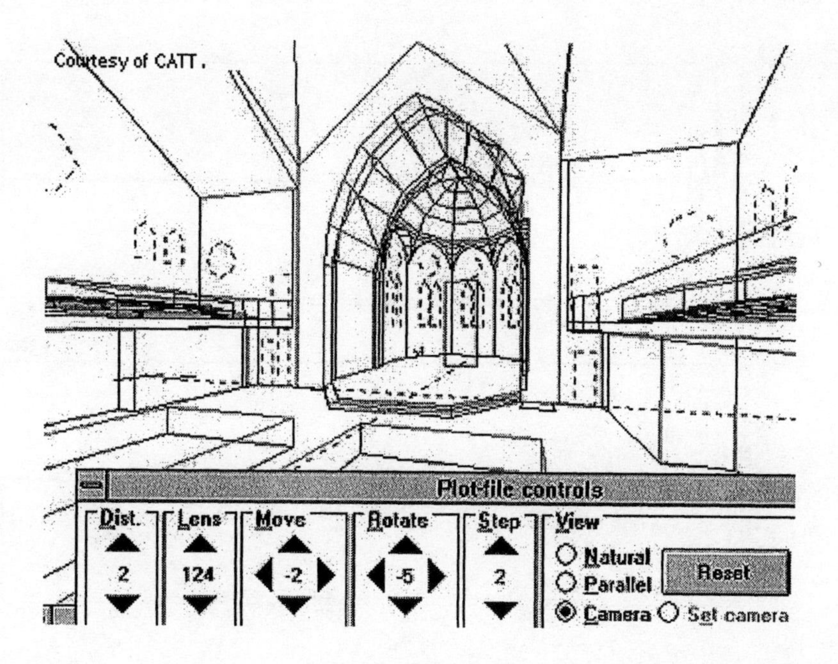

Figure 6. A view inside a church, with controls for changing view perspective.

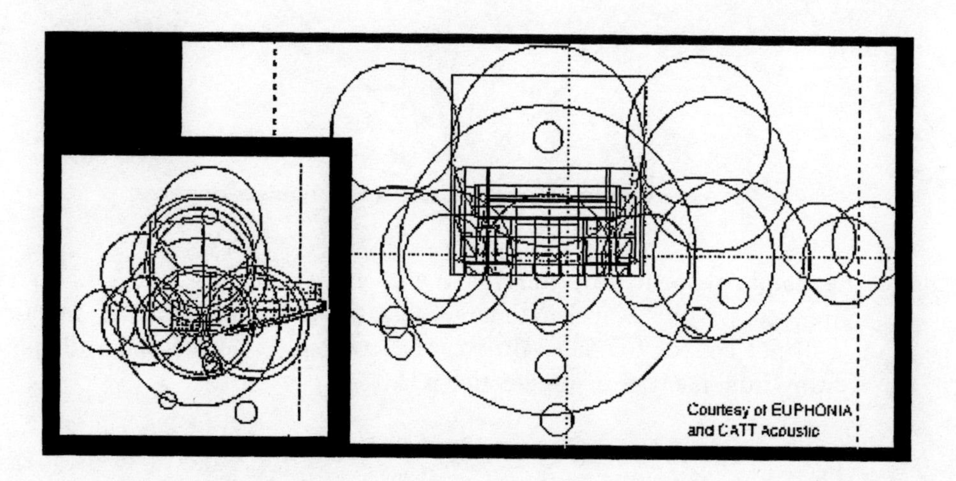

Figure 7. Spatial calculation of the intensity of virtual images. Left: side view; right: forward view. The size of the circle is the intensity of the reflection; the distance from the hall is the time delay; and the location is relative to a listener seated in the hall.

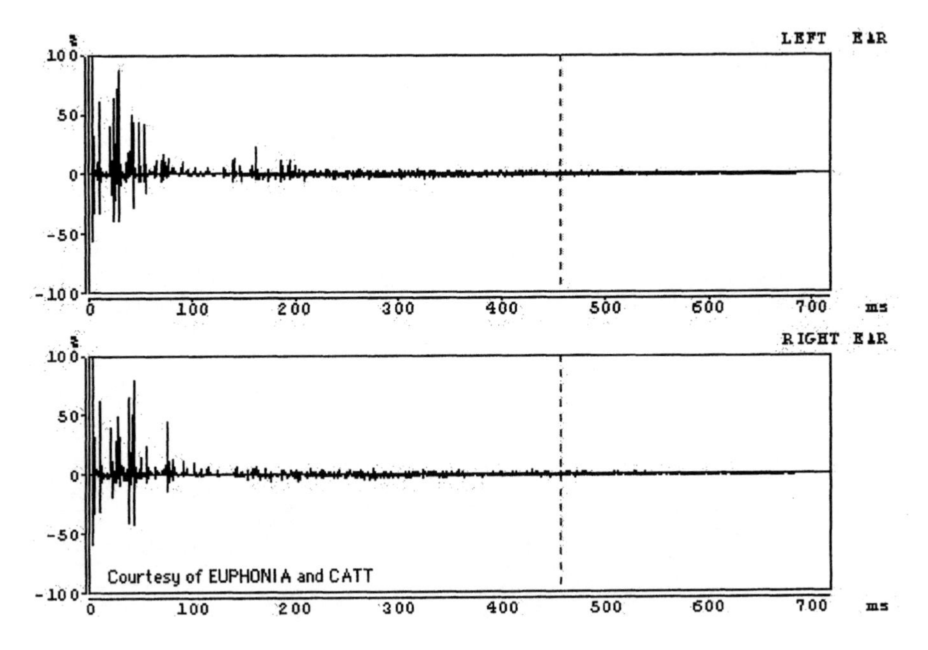

Figure 8. Binaural impulse response.

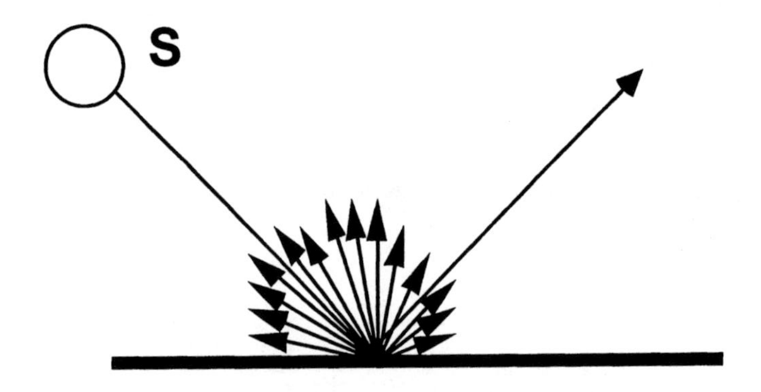

Figure 9. Example of sound diffusion from a wall, from a sound source *S* for a given frequency with a wavelength larger than the wall's surface. The waveform will not bounce off the wall like a beam of light but instead will spread away from the wall in all directions.

5 Theory

The theory and mathematics involved in modeling sound diffusion and reflection can be extremely complex; only a basic overview will be given here. One can obtain details on mathematical modeling from several sources that are relevant to auralization (e.g. Schroeder [16]; Allen and Berkley [17]; Borish [18]; Ando [15]; Lehnert and Blauert [7]; Kuttruff [19]).

The modeling of a sound wave as a ray has been used for a long time to describe the behavior of sound in an enclosure and the manner in which it reflects off a surface. This type of modeling is termed geometrical acoustics. The analogy is to light rays; one can put aluminum foil on the walls of a room, point a flashlight at a surface, and observe how the beam of light reflects off surrounding surfaces according to a pattern where the angle of incidence equals the angle of reflection. An incident sound wave behaves more or less like the light beam to the degree that its energy is specular, i.e. the wavelength is smaller than the reflective. Sound waves that are larger than the reflecting surfaces are by contrast diffused outward in a complex manner (see Figure 9). Consequently, many sounds, which are complex and contain both high and low frequencies, will be partly diffused and partly reflected.

Figure 10 shows the image model method for calculating early reflection patterns in two dimensions (see Allen and Berkley [17]; Borish [18]). In practice, a three-dimensional model is used to determine patterns from walls, ceiling and the floor. The idea is to model sound as a single specular reflection to each surface of the modeled room, as if the sound were a light beam within a "room of mirrors". By finding the virtual sound source position in a mirror image room, a vector can be drawn from the virtual source to the receiver that crosses the reflection point.

The image model can be extended to successive orders of reflections; successive bounces are calculated by extending outwards with "mirrors of mirrors", etc. Typically, auralization systems use no more than second or third order reflections, since the number increases geometrically with each successive order. The process works well for simple room models, but for complex shapes akin to real rooms re-entrant angles will be obstructed by surfaces. Rays may or may not encounter a listener placed in the model, requiring adjustment of any final calculations (see Borish [18]; Lehnert and Blauert [7]).

Another method for determining early reflections is to use a ray-tracing process (see Figure 11). In this method, the sound source is modeled as emitting sound "particles" in all directions; the particles can be approximated by rays emitted from a sound source, according to its pattern of emission (Schroeder [16]). But unlike the image model method, where a first-order reflection is calculated once to each surface, the ray-tracing method calculates an energy distribution that is applied many times to each surface. As each ray encounters a surface, it can be appropriately filtered according to angle of incidence and wall absorption.

Ray tracing has been used for some time to obtain accurate measurements of reverberation time. Unlike the image model, the effects of diffusion can be calculated more accurately with the ray-tracing method, but it is much more computationally intensive to perform because a large number of rays are usually needed for accurate results. Some auralization systems use combined image model–ray-tracing schemes. Other computationally intensive but more accurate techniques, such as boundary element modeling (BEM) or finite element modeling (FEM), may find increased use in the future. FEM and BEM techniques are particularly important for modeling the acoustics of small spaces, such as automobile compartments [1], since the modal response of such spaces must be modeled for accurate simulation of low frequencies.

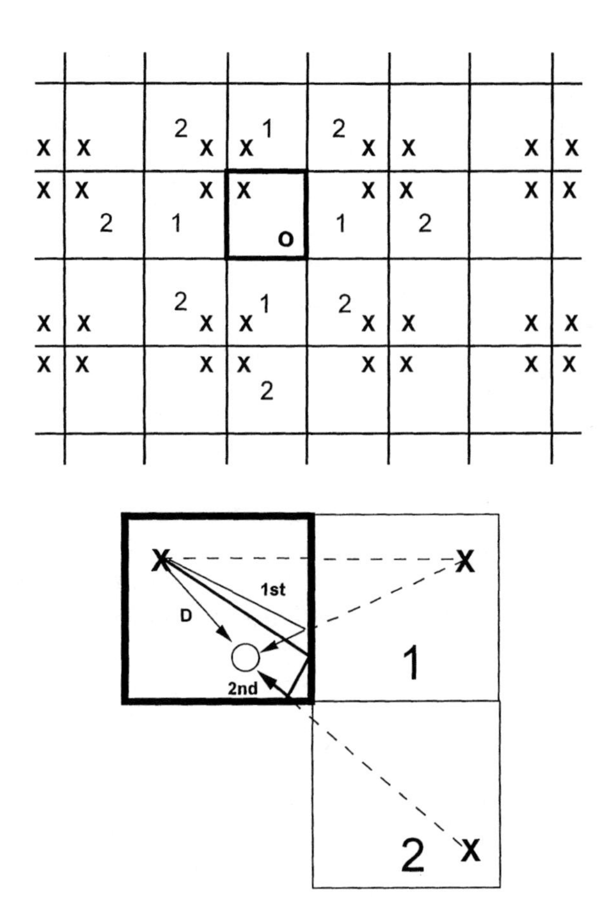

Figure 10. Use of the image model method to calculate early reflection patterns. X = location of actual and virtual sound sources; O = location of the listener. Top: A slice through the three-dimensional image space; the source in the actual room (shown with bold lines) is mirrored outward to create first-order reflections (labeled **1**); mirrors of mirrors create second-order reflections (labeled **2**), etc.. Bottom: magnification of upper illustration, showing direct sound (*D*) along with a first-order reflection (one bounce) and a second-order reflection (two bounces). Dashed lines through the real and mirrored images show the reflection point on the wall.

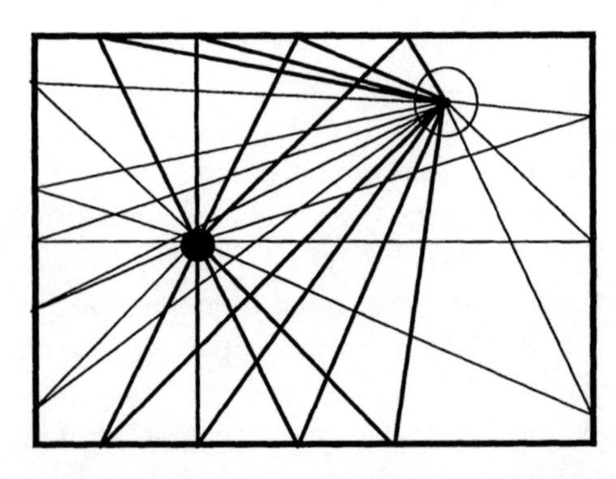

Figure 11. A simple two-dimensional ray-tracing model.

6 Perception

The application of an auralization system to a given problem will result in accurate solutions only to the degree that both acoustical behavior and human perception are modeled accurately. For instance, one study found that increasing the *number* of modeled early reflections influences the perceived spatial impression (envelopment) and distance of a sound source (Begault [20]). In another study, using information derived from an acoustical CAD program, it was found that listeners could not discriminate between different spatial incidence patterns of six HRTF-filtered "virtual early reflections" (Begault [8]). Subjects auralized the convolution of test material under three configurations. The first was facing the sound source, as derived from a room model; the second, a version with the listener turned 180 degrees; and the third, with a *random* spatial distribution of reflections. The same timings and amplitudes were used for the early reflections in each case; only the directional information contained in the HRTFs was varied. Subjects were asked to state what differences they heard between the files, in terms of timbre, spatial positioning, or loudness of the source. They were also asked to state anything else they wanted to about the sound they heard.

Because most of the subjects used were expert listeners, all reported the sensation that the sound source was in some kind of room–i.e. they noticed that reverberation was present, although only early reflections were used. However, none of the subjects reported the image switching back and forth from behind to in front of them while listening to the first and second patterns. In fact, it was extremely difficult for anyone to hear *any* difference between the three examples. The explanation given was that the spectral information in the HRTF-filtered early reflections was masked by the direct sound and adjacent reflections (see Begault [8]). The use of a more complete model of early reflections might solve the problem, but it has not yet been determined how accurate a model needs to be. Another solution may be the inclusion of head-tracked virtual acoustics, as implemented in products like the Lake HURON (Reilly and McGrath [21]).

Progress continues to be made by several researchers in specific application areas of auralization. The Room Acoustics Group at the Chalmers University of Technology (Gothenburg, Sweden) published several papers evaluating the success of auralizing actual acoustic design projects (e.g. Dalenbäck *et al.* [22]). Other studies have investigated audibility of reflections [23, 24] and edge diffraction [25]. Many other basic research projects exist as well; the sum of all of these efforts, combined with further advances in DSP hardware, should make auralization a prime 3-D audio application area in the near future.

References

[1] Kleiner, M., Dalenbäck, B. I., and Svensson, P., Auralization—an overview, *J. Audio Eng. Soc.*, **41**, pp. 861–875, 1993.

[2] Begault, D. R., *3-D Sound for Virtual Reality and Multimedia,* Academic Press Professional, Cambridge, MA, 1994.

[3] Horrall, T. R., *Auditorium acoustics simulator: form and uses,* Paper presented at the Audio Engineering Society 39th Convention, New York, 1970.

[4] Kuttruff, K. H., *Auralization of impulse responses modeled on the basis of ray-tracing results,* Paper presented at the Audio Engineering Society 91st Convention, New York, 1993.

[5] Takala, T., Hanninen, R., Valimaki, V., Savioja, L., Huopaniemi, J., Huotilainen, T., and Karjalainen, M. *An integrated system for virtual audio reality,* Paper presented at the Audio Engineering Society 100th Convention, Copenhagen, 1996.

[6] Jan, E., and Flanagan, J., *Image model for computer simulation of sound wave behavior in an enclosure,* Paper presented at the 1995 IEEE ASSP Workshop on applications of signal processing to audio and acoustics, Mohonk Mountain House, New Platz, NY, 1995.

[7] Lehnert, H., and Blauert, J., Principals of binaural room simulation, *Appl. Acoust,* **36**, pp. 259-91, 1992.

[8] Begault, D. R., *Binaural Auralization and Perceptual Veridicality,* Paper presented at the 93rd Audio Engineering Society Convention, San Francisco, 1992.

[9] Jørgensen, M., Ickler, C. B., and Jacob, K. D., *Using subject-based testing to evaluate the accuracy of an audible simulation system*, Paper presented at the 95th Audio Engineering Society Convention, New York, 1993.

[10] Kendall, G. S., and Martens, W. L., Simulating the cues of spatial hearing in natural environments, *Proceedings of the 1984 International Computer Music Conference*, International Computer Music Association, San Francisco, 1984.

[11] Foster, S. H., Wenzel, E. M., and Taylor, R. M., Real-time synthesis of complex acoustic environments (Summary), in *Proceedings of the ASSP (IEEE) Workshop on Applications of Signal Processing to Audio and Acoustics*, IEEE Press, New York, 1991.

[12] Begault, D. R., Perceptual effects of synthetic reverberation on three-dimensional audio systems, *J. Audio Eng. Soc.*, **40**, pp. 895–904, 1992.

[13] Korenaga, Y., and Ando, Y., A sound–field simulation system and its application to a seat-selection system, *J. Audio Eng. Soc.*, **41**, pp. 920–930, 1993.

[14] Ando, Y., *Concert Hall Acoustics*, Springer-Verlag, Berlin, 1985.

[15] D'Antonio, P., Konnert, J., and Kovitz, P., *Disc Project: Auralization using directional scattering coefficients,* Paper presented at the 95th Audio Engineering Society Convention, New York, 1993.

[16] Schroeder, M. R., Digital simulation of sound transmission in reverberant spaces, *J. Acoust. Soc. Am.*, **47**, pp. 424–431, 1970.

[17] Allen, J. B., and Berkley, D. A., Image model for efficiently modeling small–room acoustics, *J. Acoust. Soc. Am.*, **65**, pp. 943–950, 1979.

[18] Borish, J., *Electronic Simulation of Auditorium Acoustics,* Ph.D. Dissertation, Stanford University, 1984.

[19] Kuttruff, H., *Room Acoustics,* Elsevier Science Publishers, Essex, UK, 3rd edn., 1991.

[20] Begault, D. R., Control of auditory distance, Ph.D. Dissertation, University of California, San Diego, 1987.

[21] Reilly, A., and McGrath, D., *Convolution processing for realistic reverberation,* Paper presented at the 98th Audio Engineering Society Convention, Paris, 1995.

[22] Dalenbäck, B.-I., Kleiner, M., and Svensson, P., Audibility of changes in geometric shape, source directivity, and absorptive treatment—experiments in Auralization, *J. Audio Eng. Soc.*, **41**, pp. 905–913, 1993.

[23] Begault, D. R., *Audible and inaudible early reflections: thresholds for auralization system design,* Paper presented at the 100th Audio Engineering Society Convention, Copenhagen, 1996.

[24] Begault, D. R., Wenzel, E. M., Tran, L. L., and Anderson, M. R., *Octave-band thresholds for modeled reverberant fields,* Paper presented at the Audio Engineering Society 104th Convention, Amsterdam, 1998.

[25] Torres, R. R., On the audibility of diffraction and of frequency-response smoothing in auralization, Ph.D. Dissertation, Chalmers University of Technology, Göteberg, Sweden, 1998.

Chapter 4

Fundamental subjective attributes of sound fields based on the model of auditory-brain system

Y. Ando, S. Sato and H. Sakai
Graduate School of Science and Technology, Kobe University
Rokkodai, Nada, Kobe 657-8501 Japan
Email: andoy@kobe-u.ac.jp

Abstract

This chapter describes fundamental subjective attributes of sound fields from the model of the human auditory-brain system. The model consists of the autocorrelators and the cross-correlator for signals arriving at two ear entrances. Primary sensations such as pitch, loudness and timbre are first discussed. Then typical fundamental attributes of sound fields, including the apparent source width (ASW) and speech intelligibility, are described in terms of the spatial factors extracted from the inter-aural cross-correlation function, and the temporal factors extracted from the autocorrelation function, respectively.

1 Model

The sensitivity of a human ear to a sound source in front of the listener is essentially determined by the physical system between the source and the oval window of the cochlea (Ando [1]). Specific characteristics of the electro-physiological responses of both left and right human cerebral hemispheres (Ando *et al*. [2]; Ando *et al*. [3]; Ando *et al*. [4]; Ando [5]; Ando and Chen [6]; Chen and Ando [7]; Nishio and Ando [8]) are taken into account for the model shown in Figure 1 (Ando [9]). The sound source $p(t)$ in this Figure is located at a point r_0 in a three-dimensional space, and the listener sitting at r is defined by the location of the center of the head, $h_{l,r}(r|r_0, t)$ being the impulse responses between r_0 and the left and right ear-canal entrances. The impulse responses of the external ear canal and the bone chain are respectively $e_{l,r}(t)$ and $c_{l,r}(t)$. The velocities of the basilar membrane are expressed by $V_{l,r}(x, \omega)$, x being the position along the membrane. The action potentials from the hair cells are conducted and transmitted to the cochlear nuclei, the superior olivary complex (including the medial superior olive, the lateral superior olive and the trapezoid body), and to the higher levels of the two cerebral hemispheres.

The input power density spectrum of the cochlea $I(x')$ can be roughly mapped, according to the tuning of a single fiber (Katsuki *et al.* [10]; Kiang [11]), at a certain nerve position x'. This fact may be partially supported by ABR waves (I-IV) which reflect the sound pressure levels as a function of the horizontal angle of incidence to a listener. Such neural activities, in turn, include sufficient information to produce the autocorrelation function (ACF) at a higher level, probably near the lateral lemniscus, as indicated by $\Phi_{ll}(\sigma)$ and $\Phi_{rr}(\sigma)$, where σ corresponds to the neural activities. For convenience, the interchange of neural signals is not included here. As discussed by Ando *et al.* [2], the neural activity (wave V) may correspond to the IACC. Thus, the inter-aural cross-correlation mechanism may exist at the inferior colliculus. It is concluded that the output signal of the inter-aural cross-correlation mechanism including the IACC and the loci of maxima may be dominantly connected to the right hemisphere. The sound pressure level may be expressed by a geometrical average of ACFs for the two ears at the time of origin ($\sigma = 0$), which in fact appearing in the latency at the inferior colliculus, may be processed in the right hemisphere. Effects of the initial time delay gap between the direct sound and the single reflection Δt_1 included in the autocorrelation function may activate the left hemisphere. Such specialization of the human cerebral hemispheres may be related to the highly independent contributions of spatial and temporal criteria to subjective attributes. It is remarkable that "cocktail party effects," for example, can be well explained by such specialization of the human brain because speech is processed in the left hemisphere and spatial information is mainly processed in the right hemisphere.

Using the model, we can describe any subjective attribute of sound fields in terms of processes of the auditory pathways and the brain. The power density spectra in the neural activities in the left and right auditory pathways have a sharpening effect (Katsuki *et al.* [10]; Kiang [11]). This information is enough to attain the approximation of autocorrelation functions $\Phi_{ll}(\sigma)$ and $\Phi_{rr}(\sigma)$, respectively. Together with the mechanism of the inter-aural cross-correlation, this model can well describe fundamental subjective attributes.

Figure 1 (opposite). An auditory-brain model with the autocorrection mechanism, the inter-aural cross-correlation mechanism and the specialization of human brain related to the spatial and temporal factors for subjective responses.

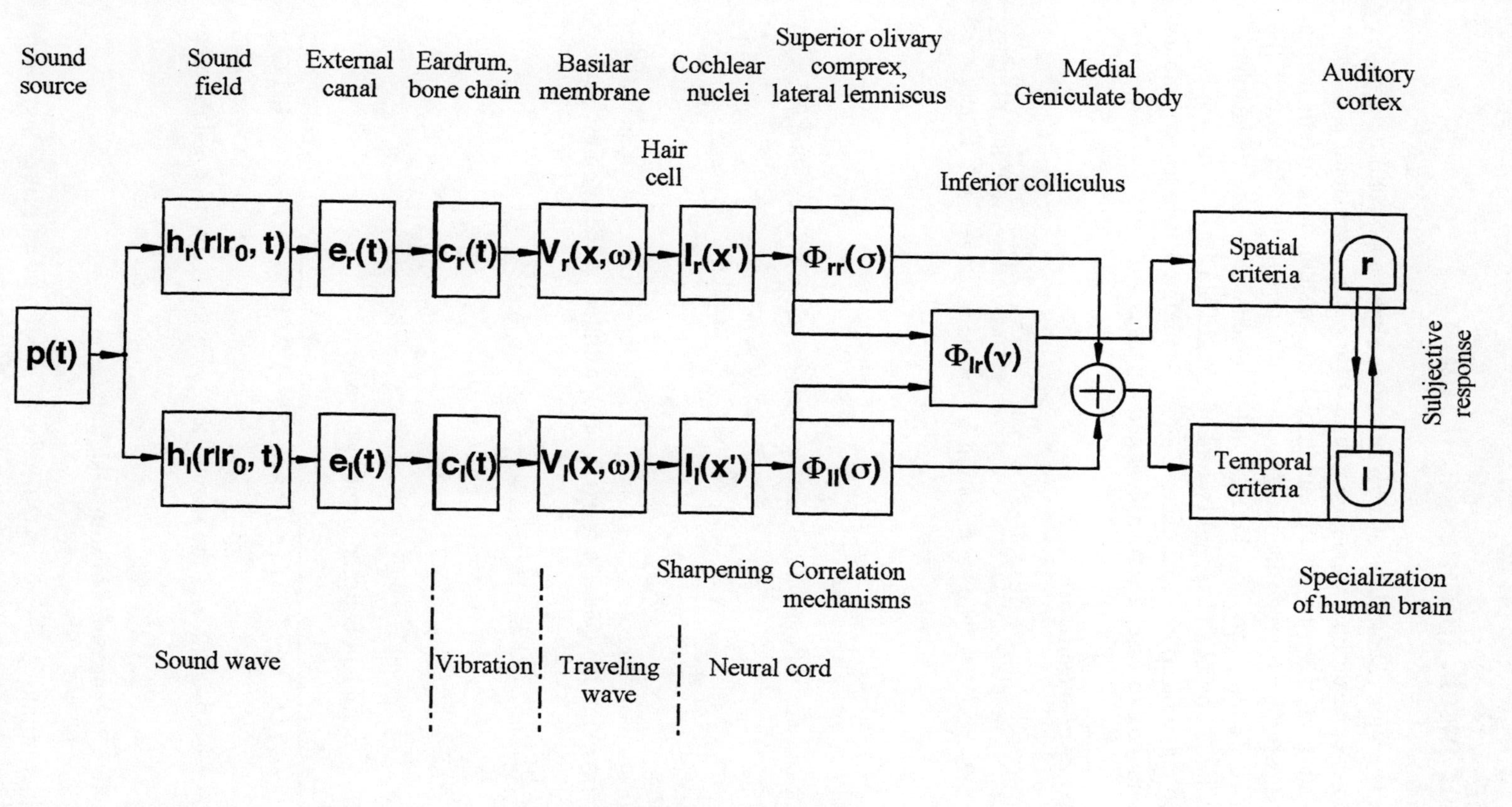

Sound source
Sound field
External canal
Eardrum, bone chain
Basilar membrane
Cochlear nuclei
Superior olivary comprex, lateral lemniscus
Medial Geniculate body
Auditory cortex
Hair cell
Inferior colliculus
p(t)
h_r(r|r_0, t)
e_r(t)
c_r(t)
V_r(x,ω)
I_r(x')
Φ_rr(σ)
Φ_lr(ν)
Spatial criteria
r
h_l(r|r_0, t)
e_l(t)
c_l(t)
V_l(x,ω)
I_l(x')
Φ_ll(σ)
Temporal criteria
l
Subjective response
Sharpening
Correlation mechanisms
Specialization of human brain
Sound wave
Vibration
Traveling wave
Neural cord

Figure 1.

2 Theory for pitch, loudness and timbre

A description of the primary sensations of a given source signal and sound field–pitch, loudness and timbre–is based on a model consisting of the autocorrelation and interaural cross-correlation mechanisms of the auditory-brain system. It is worth noting that the power density spectrum is identical to the autocorrelation function of the signal. To describe timbre, we should take into consideration the hemispheric specialization for the temporal and spatial factors of sound fields by a similar manner to the description of subjective preference. Timbre is defined here as a primary sensation that cannot be expressed by only pitch and loudness.

2.1 A general theory of sensations

Let c_i ($i = 1, 2, ..., I$) be physical factors representing cues influencing any primary sensation s_j ($j = 1, 2, ..., J < I$). Then a sensation s_j may be expressed as a function of those factors:

$$s_j = f(c_1, c_2, ..., c_i), \quad j = 1, 2, ..., J \tag{1}$$

If the physical factors are orthogonal, s_j may be expressed by a linear combination such that

$$s_j = f(c_1) + f(c_2) + ... + f(c_i), \quad j = 1, 2, ..., J \tag{2}$$

Here arises the question of whether or not sensation is independent of other sensations. For example, the answer to the question is easily demonstrated by the simplest case:

$$\begin{aligned} s_j &= f_j(c_1) + f_j(c_2) \\ s_k &= f_k(c_1) + f_k(c_2) \end{aligned} \tag{3}$$

Then the correlation coefficient between s_j and s_k is given by

$$r_{jk} = \overline{s_j s_k} = \overline{f_j(c_1) f_k(c_1)} + \overline{f_j(c_2) f_k(c_2)} + \overline{f_j(c_1) f_k(c_2)} + \overline{f_j(c_2) f_k(c_1)} \tag{4}$$

and is not zero because the first and the second terms on the right-hand side are not always zero.

2.2 Theory of primary sensations of source signals

Let us now consider primitive sensations (Roederer [12]) of a given source signal $p(t)$ located in front of a listener in a free field. The ACF is defined by

$$\Phi_{p}(\tau)=\lim_{T\to\infty}\frac{1}{2T}\int_{-T}^{+T}p'(t)p'(t+\tau)\mathrm{d}t \tag{5}$$

where $p'(t) = p(t)*s(t)$, $s(t)$ being the ear sensitivity. For convenience, $s(t)$ may be chosen as the impulse response of an A-weighted network.

The ACF is identical to the power density spectrum $P_d(\omega)$, so

$$\Phi_{p}(\tau)=\int_{-\infty}^{+\infty}P_{d}(\omega)e^{j\omega t}\mathrm{d}\omega \tag{6}$$

and

$$P_{d}(\omega)=\int_{-\infty}^{+\infty}\Phi_{p}(\tau)e^{-j\omega t}\mathrm{d}\tau \tag{7}$$

Thus, the ACF and the power density spectrum mathematically contain the same information. In ACF analysis, there are three significant parameters:

1. Energy represented at the origin of the delay, $\Phi_p(0)$
2. Effective duration of the envelope of the normalized ACF, τ_e (which is defined by the ten-percentile delay), representing a kind of repetitive feature or reverberation containing the source signal itself as shown in Figure 2(a). The normalized ACF (NACF) is defined by

$$\phi_p(\tau) = \Phi_p(\tau) / \Phi_p(0) \tag{8}$$

3. Fine structure, including peaks and dips with their delays. The delay time and amplitude of the first peak–namely, τ_1, and ϕ_1 as shown in Figure 2(b)–may represent this structure. These values τ_1 and ϕ_1 are usually thought to be closely related to τ_n and ϕ_n ($n > 1$), which are the delay time and amplitude of the nth peak.

Let us now consider primary sensations. Loudness s_l is given, after Ando [9], by

$$s_l = f_l(\Phi_p(0), \tau_e) \tag{9}$$

When $p'(t)$ is measured with reference to the pressure 20 μPa leading to the level $L(t)$, the equivalent sound pressure level L_{eq}, defined by

$$L_{eq}=10\log\frac{1}{T}\int_{0}^{T}10^{\frac{L(t)}{10}}\mathrm{d}t \tag{10}$$

corresponds to 10 log $\Phi_p(0)$.

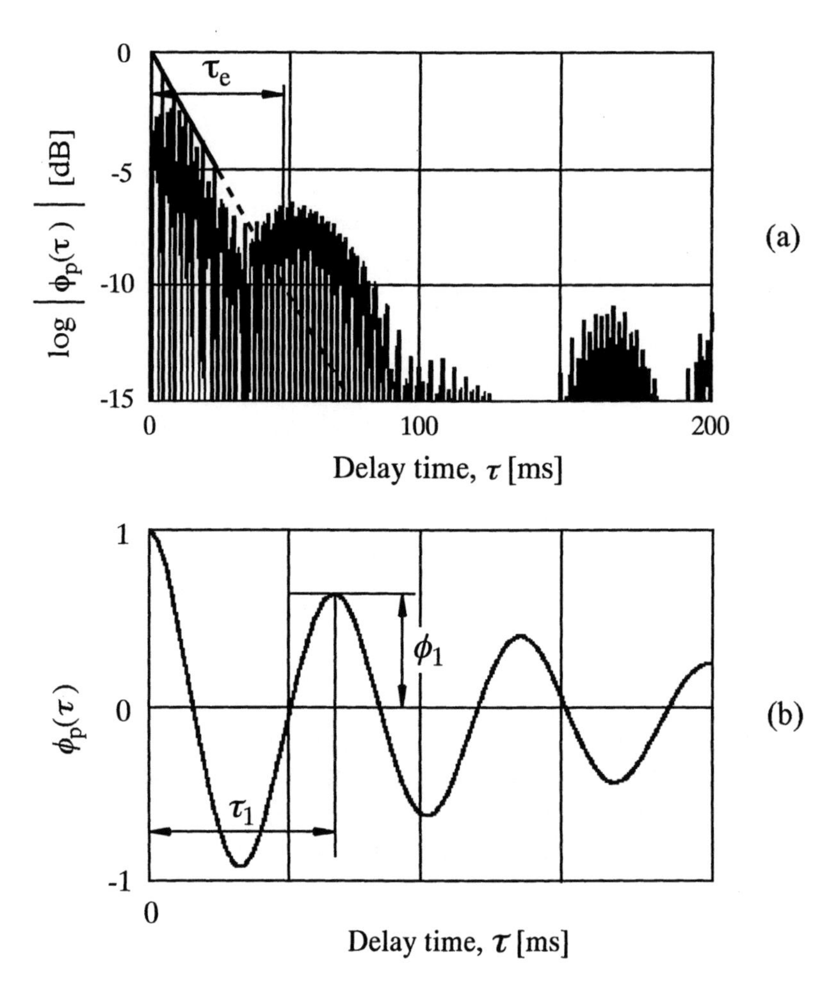

Figure 2. Definitions of independent factors other than $\Phi(0)$, extracted from the NACF, τ_e defined by the ten-percentile delay (at −10 dB), obtained practically from the decay rate extrapolated in the range from 0 dB to −5 dB of NACF (a); and τ_1 and ϕ_1 in the fine structure of the NACF (b).

A good example of applying the ACF is in the discussion of the missing fundamentals of music signals. When the signal contains only a number of harmonics without the fundamental frequency, we hear the fundamental as a pitch (Wightman [13]). As discussed in Section 4.1, this phenomenon is well explained by the fine structure of the ACF.

Pitch s_p can be expressed by

$$s_p = f_p(\tau_1, \phi_1) \tag{11}$$

where the factor ϕ_1 is the strength of pitch.

The most complicated sensation is timbre s_t defined as a remaining primitive sensation that cannot be expressed by both well-defined pitch and loudness:

$$s_t = f_t[\Phi(0), \tau_e, \tau_1, \phi_1] \tag{12}$$

2.3 Theory of primary sensation and primitive response for sound fields

There are four orthogonal factors of sound fields: listening level LL, initial time delay gap between the direct sound and the first reflection Δt_1, subsequent reverberation time T_{sub}, and the IACC (see Appendix). The IACC is related to the subjective diffuseness (Ando [1]). In addition to the IACC, the factors τ_{IACC} and W_{IACC} (Figure 3) are related to the space-oriented attributes, for example, the image shift and the ASW, respectively. Thus, all of these factors are taken into consideration when we describe the primitive sensations.

The loudness s_l of the sound field given by eqn (9) can be replaced by

$$s_l = f_l(\mathrm{LL}, \tau_e; T_{sub}) \tag{13}$$

where $\Phi(0)$ in eqn (9) is replaced by LL, which is defined by eqn (40). A typical example of loudness is shown in Figure 4 as a function of log τ_e. Since τ_e depends on T_{sub}, loudness is almost determined by both LL and τ_e. It is remarkable that the early latencies P_1, N_1, and P_2 of slow vertex responses for a short speech signal decrease with increasing sensation level (Figure 5).

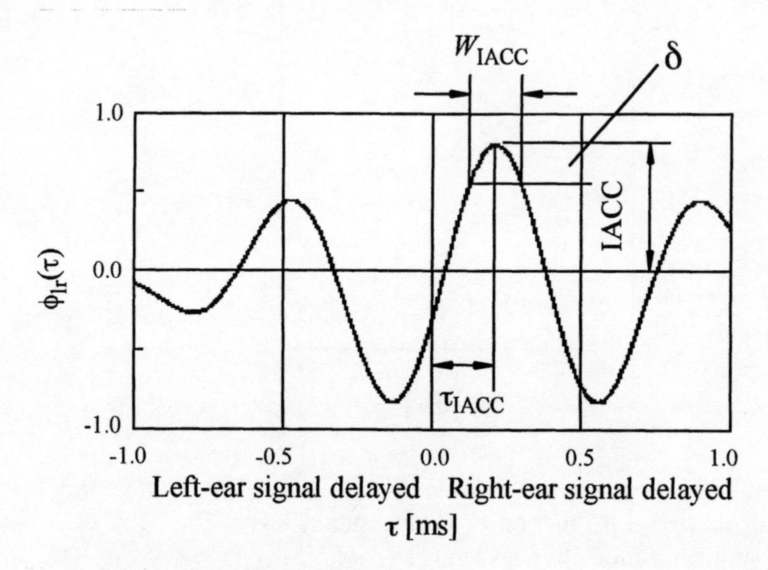

Figure 3. Definitions of independent factors extracted from the normalized inter-aural cross-correlation function, IACC, τ_{IACC} and W_{IACC}.

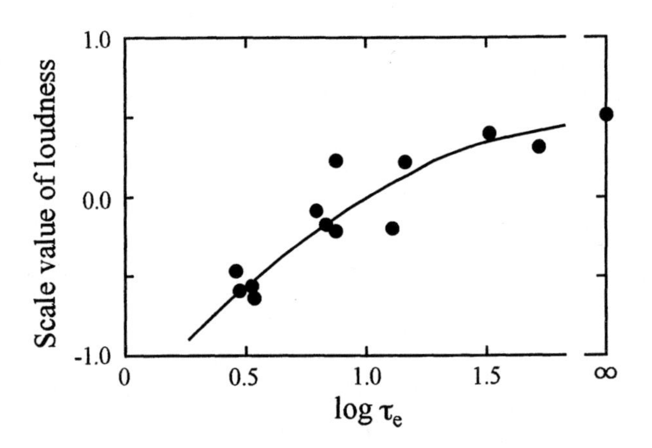

Figure 4. Scale values of loudness obtained by the paired-comparison tests, as a function of log τ_e.

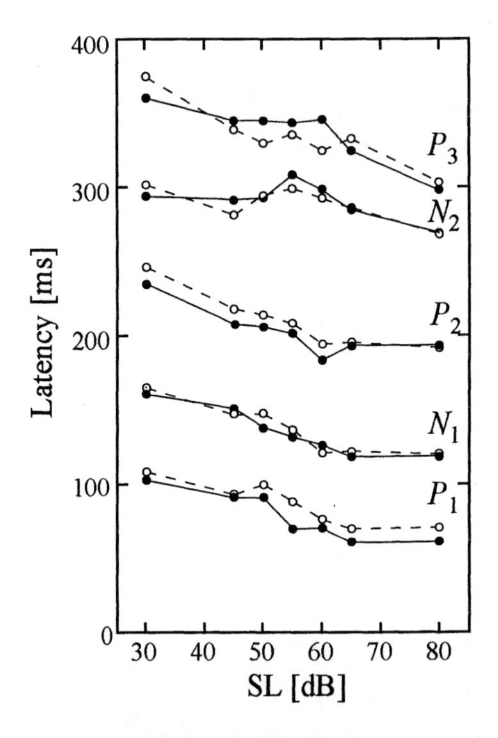

Figure 5. Latencies of slow vertex responses P_1, N_1, and P_2 for a short (0.9 s) speech signal plotted as a function of the sensation level. The peak of N_2 latency corresponds to subjective preference (Ando [5]).

The factor Δt_1 is thought to be a minor factor influencing τ_1 and ϕ_1, so pitch can be determined by only τ_1 and ϕ_1. Therefore, the pitch s_p given by eqn (11) holds for sound fields as well:

$$s_p = f_p\,(\tau_1, \phi_1) \tag{14}$$

Timbre may be expressed by all of the orthogonal factors of both the sound signal and the sound field, so

$$s_t = f_t[\tau_e, \tau_1, \phi_1; \mathrm{LL}, \Delta t_1, T_{sub}, \mathrm{IACC}, \tau_{IACC}, W_{IACC}] \tag{15}$$

Considering the specialization of the human cerebral hemisphere for the temporal factors of the left hemisphere and special factors of the right hemisphere, we can rewrite eqn (15) as

$$s_t = f_t[\tau_e, \tau_1, \phi_1; \mathrm{LL}, \Delta t_1, T_{sub}]_{left} + f_t[\mathrm{IACC}, \tau_{IACC}, W_{IACC}]_{right} \tag{16}$$

For example, coloration of the single reflection for sound field is described by the left hemispheric factors, namely Δt_1 and the τ_e of the ACF, as demonstrated in Figure 6 (Ando and Alrutz [14]). Timbre is considered an overall subjective response similar to the subjective preference. Timbre might be investigated in referring to the theory of primitive–subjective preference (Ando [1]).

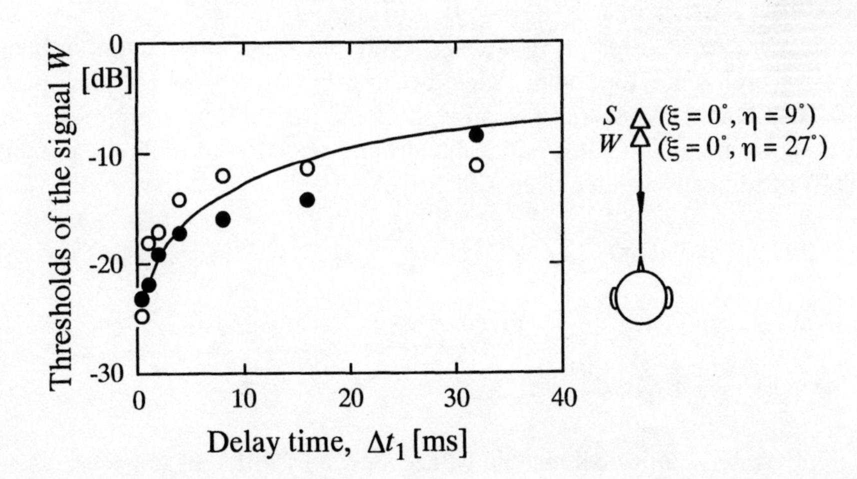

Figure 6. Coloration thresholds of the single weak reflection of the frontal incidence as a function of its delay time Δt_1 (Ando and Alrutz [14]). Dashed curve is obtained by the ACF-envelope of the source signal of the band-limited noise centered at 1 kHz.

It is quite remarkable that information corresponding to primitive–subjective preference of sound fields is found in specific components of brain waves. The latency of the slow vertex responses (Ando [5]) and the effective duration of ACF of the α-wave frequency range of continuous brain waves well correspond to the scale values of subjective preference (Ando [9]). The scale value of subjective preference is described by

$$s = f(\Delta t_1/[\Delta t_1]_p, T_{sub}/[T_{sub}]_p)_{left} + f(LL/[LL]_p, IACC)_{right}$$
$$\approx f(\Delta t_1/[\Delta t_1]_p) + f(T_{sub}/[T_{sub}]_p) + f(LL/[LL]_p) + f(IACC) \quad (17)$$

This formula holds when $\tau_{IACC} = 0$, and the suffix p indicates the most preferred value of each factor. The most preferred conditions of the temporal factors for a number of subjects are approximately given by

$$[\Delta t_1]_p \approx (1\text{-}\log_{10} A)\, \tau_e \quad (18)$$

$$[T_{sub}]_p \approx 23\, \tau_e \quad (19)$$

and

$$A = \left(\sum_{n=1}^{\infty} A_n^2 \right)^{1/2} \quad (20)$$

where A is the total amplitude of reflections and A_n is an amplitude of the nth reflection. A smaller value of the IACC is better for all the subjects investigated (Ando [9]).

It is interesting that τ_1, ϕ_1 and W_{IACC} are not significant factors influencing the subjective preference for sound fields, because they are related to the source signal itself. The timbre is an overall response similar to the subjective preference. It is quite natural to assume that the just noticeable difference (JND) of timbre is proportional to the JND of subjective preference (Ando [9]), so

$$JND(s_t) = \kappa JND(s) \quad (21)$$

The JND for each factor may be given by

$$JND(s_i) = \kappa_i\, \partial s_i/\partial c_i, \quad i = 1, 2, \ldots, J \quad (22)$$

Different symbols indicate threshold values judged by different subjects.
where κ_i is constant and s_i is given, after Ando [9], by

$$s_i \approx -\alpha_i |x_i|^{3/2} \quad (23)$$

where x_i is a physical parameter and α_i is a weighting coefficient of subjective preference. It is worth noting that, in the preferred range of each factor c_i, s_i is rather flat, so that it might be hard to judge differences in timbre as well as subjective preference.

Timbre has so far been discussed here as an overall response, and it might be well described in terms of the model of the auditory-brain system. The subjective preference is described as a primitive response in Ando [9], but it is not treated here.

3 Spatial subjective attributes in relation to spatial factors extracted from the inter-aural cross-correlation function

Independent factors extracted from the inter-aural cross-correlation function are defined in Figure 2. The τ_{IACC} a significant factor in determining the perceived horizontal direction of the sound or image shift, and the IACC is the degree of subjective diffuseness of the sound field (Damaske and Ando [15]). A well-defined direction is perceived when the normalized inter-aural cross-correlation function has one sharp maximum, a high value of the IACC, and a narrow value of the W_{IACC}. On the other hand, subjective diffuseness or no spatial directional impression corresponds to a low value of IACC (< 0.15). The subjective diffuseness or spatial impression of the sound field in a room is one of the fundamental attributes in describing good acoustics. It is worth noting that if the sounds arriving at the two ears are dissimilar (IACC = 0), then different signals (but signals containing the same information) are conveyed through the two independent auditory channels to the brain (Nakajima and Ando [16]).

3.1 Subjective diffuseness

The scale value of subjective diffuseness was obtained in paired-comparison tests with bandpass Gaussian noise by varying the horizontal angle of two symmetric reflections (Ando and Kurihara [17]; Singh *et al.* [18]). Listeners judged which of two sound fields was perceived as the more diffuse. A remarkable finding is that the scale values of subjective diffuseness are inversely proportional to the IACC (Figure 7) and may be formulated in terms of the 3/2 power of the IACC in a manner similar to that in which the subjective preference is formulated:

$$S \approx -\alpha(\text{IACC})^{\beta}, \tag{24}$$

where $\alpha = 2.9$ and $\beta = 3/2$.

The results of scale values obtained in the paired-comparison test and values calculated by eqn (24) are shown in Figure 8 as a function of the IACC. There is a great variation of data when IACC < 0.5, but there are no essential differences between the results obtained at frequencies between 250 Hz and 4 kHz. The scale values of subjective diffuseness, which depend on the horizontal angle, are shown in

Figure 7 for 1/3 octave-bandpass noise with a center frequency of 1 kHz. The most effective horizontal angles of reflections differ depending on the frequency range and are inversely related to the IACC values. As demonstrated in Figure 7, these are about ±55° for the 1 kHz range.

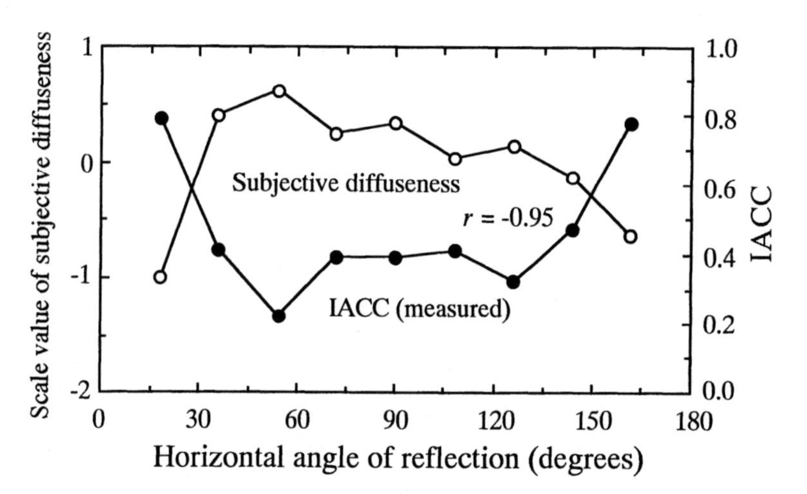

Figure 7. Scale values of subjective diffuseness and the IACC as a function of the horizontal angle of incidence to a listener. These are values obtained with 1/3 octave-bandpass noise with a center frequency of 1 kHz (seven subjects).

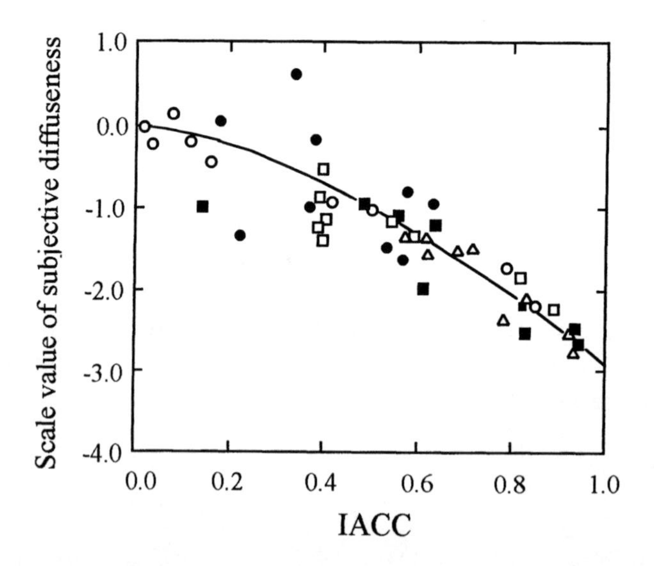

Figure 8. Scale values of subjective diffuseness as a function of the IACC calculated. Different symbols indicate different frequencies of the 1/3 octave-bandpass noise: (△) 250 Hz, (○) 500 Hz, (□) 1 kHz, (●) 2 kHz, (■) 4 kHz. (—) Regression line by eqn (24).

3.2 Apparent source width (ASW)

For a sound field with a predominately low frequency range, the inter-aural cross-correlation function has no sharp peaks for the delay range of $|\tau| < 1$ ms, and W_{IACC} becomes wider (Figure 9). The values of W_{IACC} for bandpass noise are calculated by using the following equation (Ando [9]):

$$W_{IACC}^{(\delta)} \approx \frac{4}{\Delta\omega_c}\cos^{-1}\left(1-\frac{\delta}{IACC}\right) \text{ [s]} \tag{25}$$

where $\Delta\omega_c = 2\pi(f_1+f_2)$, and f_1 and f_2 are the lower and upper frequencies of an ideal filter. For simplicity, δ is defined by 0.1(IACC).

Of particular interest is that a wider ASW may be perceived within the low frequency bands and by decreasing the IACC. More clearly, the ASW may be well described by both IACC and W_{IACC} (Sato and Ando [19]). The scale values of ASW were obtained by paired comparison tests with ten subjects. The values of W_{IACC} were varied by changing the center frequencies of 1/3 octave-bandpass noises. The center frequencies were 250 Hz, 500 Hz, 1 kHz and 2 kHz. The values of IACC were adjusted by controlling the ratio of the sound pressure of the reflections to sound pressure of the direct sound. Because the listening level affects ASW (Keet [20]), the total sound pressure levels at the ear canal entrances were kept constant at a peak of 75 dBA. Subjects judged which of two sound sources they perceived to be wider. The results of the analysis of variance for the scale values s(ASW) indicate that both the factors IACC and W_{IACC} are significant ($p < 0.01$) and contribute to the s(ASW) independently such that

$$\begin{aligned} s(\text{ASW}) &= f(\text{IACC}) + f(W_{IACC}) \\ &\approx a(\text{IACC})^{3/2} + b(W_{IACC})^{1/2} \end{aligned} \tag{26}$$

where coefficients $a \approx -1.64$ and $b \approx 2.44$ are obtained by regressions of the scale values with ten subjects as shown in Figures 10(a) and 10(b). As shown in Figure 11, scale values s(ASW) calculated by eqn (26) and measured scale values are obviously in good agreement ($r = 0.97, p < 0.01$).

Scale values s(ASW) for each listener can also be calculated by the same equation used in calculating the global s(ASW). Coefficients a and b in eqn (26) for each subject were calculated by a multiple regression analysis and are listed in Table 1. Average values of coefficients a and b for all subjects correspond to the values of global result. Figure 12 shows the relation between the measured scale values s(ASW) obtained from the test and the scale values calculated by eqn (26) for each subject. The different symbols indicate the different subjects. The correlation coefficient between the measured and calculated s(ASW) is 0.91 ($p < 0.01$). The ASW for each subject can be obtained by using the same equation used in calculating the global ASW simply by changing the weighting coefficients a and b.

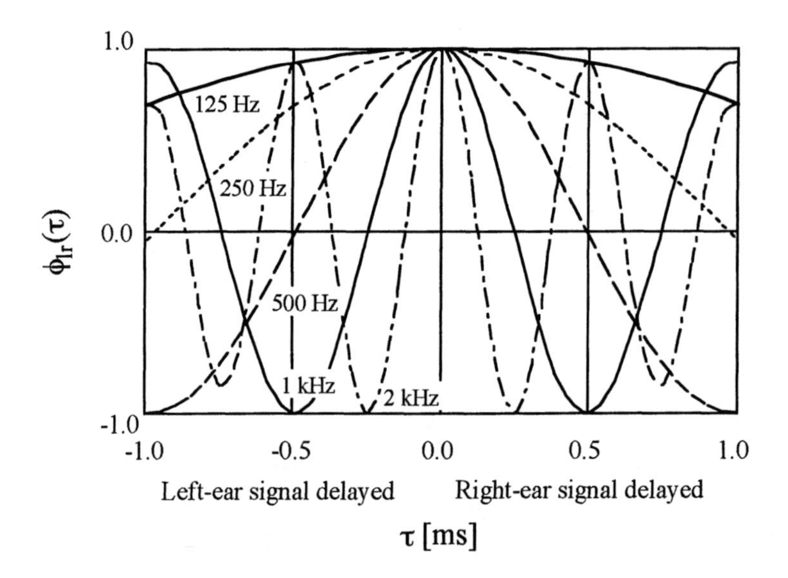

Figure 9. Measured inter-aural cross-correlation functions for 1/3 octave-bandpass noises with center frequencies of 125 Hz to 2 kHz.

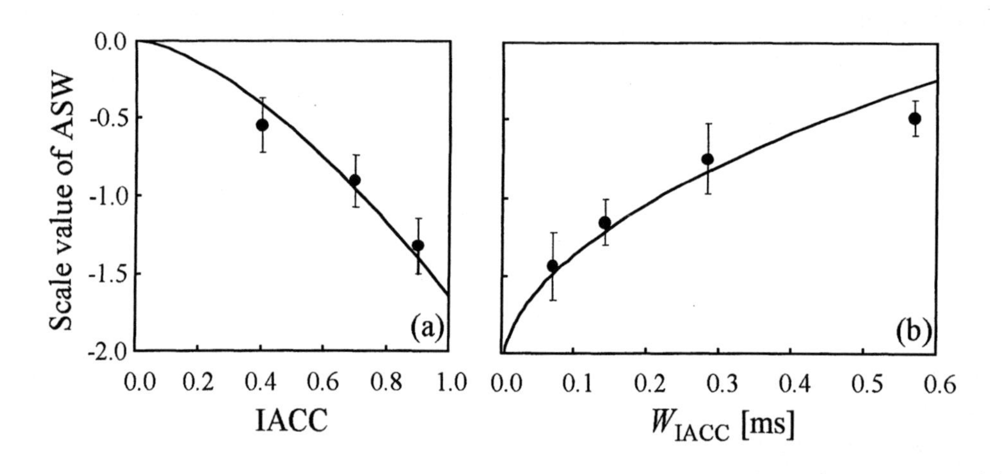

Figure 10. Average scale values of ASW for 1/3 octave-bandpass noises with 95% reliability, as a function of the IACC (a). The regression curve is expressed by the first term of eqn (26) with $a \approx -1.64$. Average scale values of ASW for 1/3 octave-bandpass noises with 95% reliability, as a function of W_{IACC} (b). The regression curve is expressed by the second term of eqn (26) with $b \approx 2.44$.

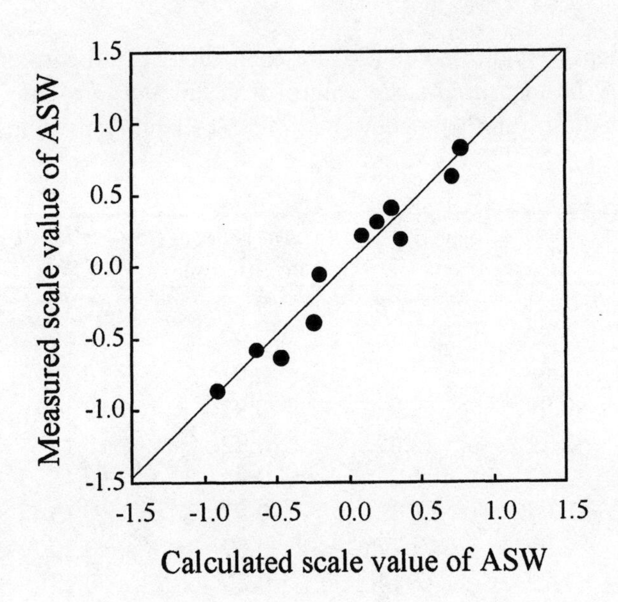

Figure 11. Relation between the measured scale values of ASW and the scale values s(ASW) calculated by eqn (26) ($r = 0.97, p < 0.01$).

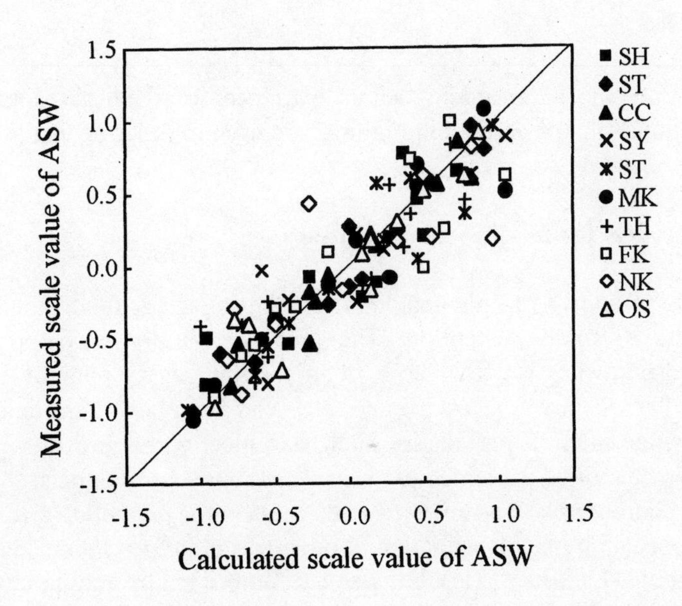

Figure 12. Relation between the measured scale values for 1/3 octave-bandpass noises and scale values calculated by eqn (26) for each subject ($r = 0.91, p < 0.01$).

Table 1. Coefficients a and b in eqn (26) for each subject, and correlation coefficients between the measured scale values of ASW and the scale values of ASW calculated by eqn (26) together with 95% reliability and a maximum error of SV.

Individual	a	b	Correlation coefficient	95% reliability of SV	Maximum errorof SV
SH	−1.21	2.58	0.88	±0.18	0.46
TS	−1.50	3.18	0.97	±0.12	0.27
CC	−1.05	2.82	0.97	±0.10	0.21
SY	−0.94	2.92	0.91	±0.17	0.56
MK	−2.21	2.09	0.92	±0.17	0.42
ST	−2.57	1.94	0.94	±0.16	0.54
TH	−2.04	1.32	0.87	±0.19	0.59
FK	−0.99	3.27	0.89	±0.20	0.52
NK	−1.79	2.14	0.80	±0.23	0.78
OS	−2.09	2.14	0.94	±0.15	0.40
Mean ± SD	−1.64 ± 0.58	2.44 ± 0.62			

4 Subjective attributes in relation to temporal factors extracted from the autocorrelation function of sound signals

In this section, the independent factors extracted from the autocorrelation function (ACF) of sound signals defined in Figure 2 are used to describe a missing fundamental and loudness.

4.1 ACF models for missing fundamental

As is widely known, a phenomenon of the missing fundamental represents a characteristic of pitch perception. The pitch of harmonic components without a fundamental frequency is perceived as being the same as the pitch of a pure tone of the fundamental frequency. The frequency perceived from such harmonic components is called a residue pitch, a periodicity pitch, a subjective pitch or a virtual pitch. This phenomenon cannot be explained by the spectrum of source signals.

Some autocorrelation function-based models for predicting a residue pitch have been proposed. The autocorrelation model of pitch perception was originally a "duplex" model (Licklider [21]), but was later improved by adding new rules (Meddis and Hewitt [22, 23]). Three famous models called "pattern recognition" models have been generally accepted since the 1970s (Wightman [24]; Goldstein [25]; Terhardt [26]). Because the phase relation does not influence the pitch perception, a pattern-transformation model was proposed (Wightman [13, 24]). A pitch transformer based on an ACF detects the locations of peaks from the output waveform in each frequency band. Identically, the pitch strength can be estimated from the height of the maximum peak extracted from the ACF form. The effectiveness of the pattern-transformation model was examined in order to evaluate the validity in the peripheral weighting model using ripple noises (Yost et al. [27]; Yost [28, 29]). The time delay of maximum peaks of an ACF for a sound source was used as a significant parameter for predicting the pitch in their pitch-matching study. A cancellation model involving an array of delay lines and inhibitory gating neurons has also been proposed as an extension of the autocorrelation model (Cheveigne [30]). It seems reasonable to suppose, as Ohgushi [31] proposed, that both spatial and temporal cues relate to the perception of a residue.

Here, the fine structures of ACF, ϕ_1 and τ_1 in Figure 2(b) based on the human auditory-brain system are fairly adopted to predict a residue pitch of complex tones and complex noises. Results of two pitch identification tests are presented here.

4.2 Pitch perception for complex tones

The pitch-matching tests were performed by comparing pitches of complex tones and a pure tone (Sumioka and Ando [32]; Sakai *et al.* [33]). The test signals were the complex tones consisting of the third to seventh harmonics of the fundamental frequency of 200 Hz, and all partial tones had the same amplitudes as shown in Figure 13(a). As test signals, the two waveforms of complex tones, (i) in-phases and (ii) random-phases, were applied as shown in Figures 13(b) and 13(c). The starting phases of all components of the in-phase stimuli were adjusted at zero. The phases of the components of random-phase stimuli were randomly set to avoid having any periodic peak in its real waveform. As shown in Figure 13(d), the normalized ACFs (NACF) of these stimuli were calculated at the integrated interval $2T = 0.8$ s. Although the waveforms differ greatly from each other, as shown in Figures 13(b) and 13(c), their NACFs are identical. For example, the time delay at the first maximum peak of the NACF, τ_1, is 5 ms (200 Hz), which corresponds to pitch. The subjects were five musicians (two male and three female, 20–26 years of age). Test signals were radiated from a loudspeaker in front of each subject in a semi-anechoic chamber. The sound pressure level (SPL) of each complex tone at the center position of the listener's head was fixed at 74 dB by analysis of the ACF $\Phi(0)$. The distance between a subject and the loudspeaker was 0.80±0.01 m.

The probability of matching frequencies counted for each 1/12-octave band (chromatic scale) of the in-phase stimuli and random-phase stimuli are shown in Figures 14(a) and 14(b). The dominant pitch of 200 Hz is included neither in the spectrum nor in the real waveform of random phases, but it is obviously included in the period in the ACF (Figure 13). For both in-phase and random-phase conditions, about 60% of the responses are clustered within a semitone of the fundamental. Obviously, there are no fundamental differences in the distributions of pitch-matching data between the two conditions. For more detail, histograms between 183.3 Hz and 218.0 Hz for all the subjects are shown in Figures 14(c) and 14(d) for in-phase stimuli and random-phase stimuli. Averaged values and standard deviations (SD) of the data obtained from each subject at frequencies near 200 Hz are listed in Table 2. The results obtained for the pitch under the two conditions are definitely similar. In fact, the pitch strength remains the same under both conditions. Thus, the pitch of complex tones can be predicted from the time delay at the first maximum peak of the NACF, τ_1.

Table 2. Mean and standard deviation (SD) of pitch-matching data for each subject.

	Mean, Hz		SD	
Subject	In-phase	Random-phase	In-phase	Random-phase
A	202.6	201.0	1.89	2.44
B	199.1	198.3	1.70	1.42
C	202.5	202.1	1.18	1.76
D	203.7	201.7	2.29	1.65
E	202.2	202.2	1.87	2.07
Total	201.9	201.0	2.43	2.38

Individual differences in pitch perception were also found. The results for each subject are indicated in Figures 15(a) through 15(e). Subjects B and D matched only around the fundamental frequency (200 Hz). As shown in Figures 14(a) and 14(b), about 20% of the responses were clustered around 400 Hz, and the NACF has a distinct dip at $\tau = 2.5$ ms. But an octave shift for a phase change (Lundeen and Small [34]) was not observed in the results obtained from these subjects. Subjects A and E matched at the fundamental frequency and the frequency an octave higher. This octave change is thought to be caused by a similarity for the octave relation. The time delay of the ACF for this pitch is 2.5 ms, so this pitch cannot be predicted because of a dip in the ACF structure. None of the subjects matched at $\tau = 10$ ms (100 Hz), which is an octave lower than the fundamental frequency, though there is a peak at $\tau = 10$ ms. Subject C matched in three categories of center frequencies (200.0 Hz, 224.5 Hz and 317.5 Hz). An E-flat note corresponds to a center frequency of 317.5 Hz and a G note corresponds to 200.0 Hz. This is a harmonized relation (E flat: G = do : mi). Subject C seemed to seek such a harmonized relation because he is a musician who uses the key of E-flat. The results for Subject C influenced the histograms for the global results that appeared in Figures 14(a) and 14(b).

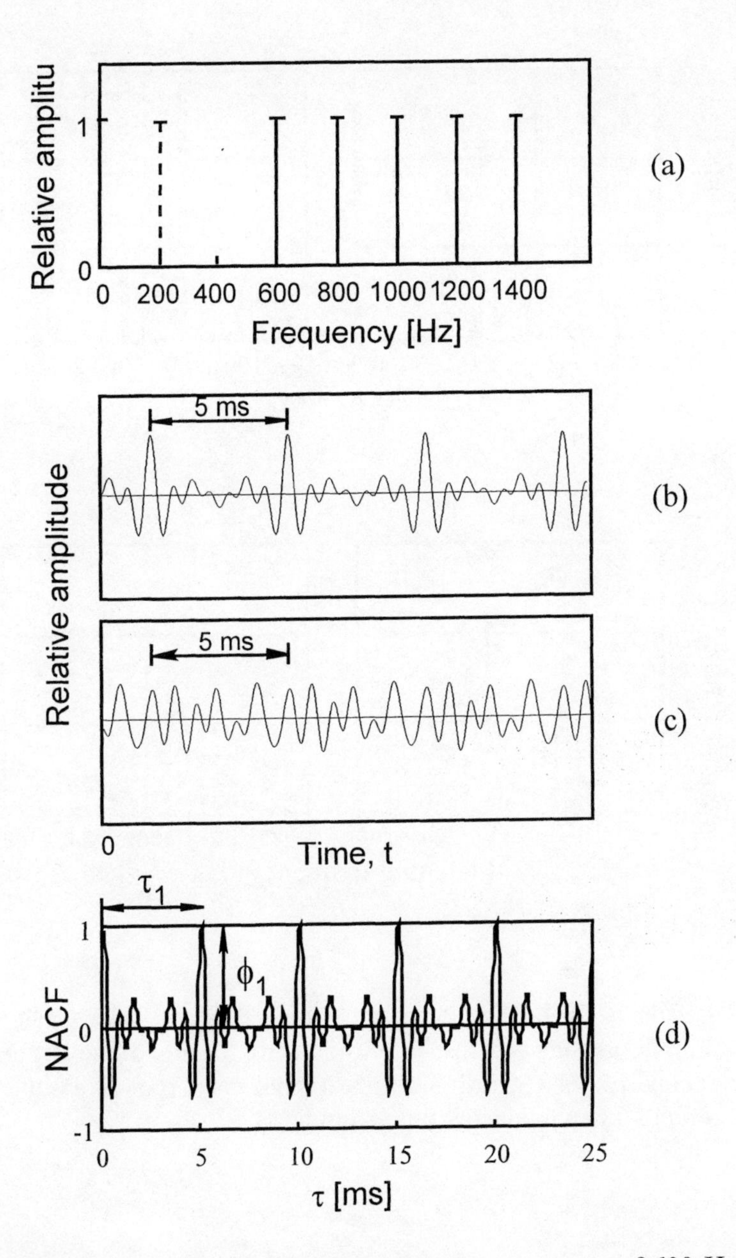

Figure 13. Complex tones presented with pure-tone components of 600 Hz, 800 Hz, 1000 Hz, 1200 Hz and 1400 Hz without the fundamental frequency of 200 Hz (a). Real waveform of complex tones in phase components (b). Real waveform of random phase components (c). Normalized autocorrelation function (NACF) of the two complex tones with its period of 5 ms (200 Hz) (d). Both τ_1 and ϕ_1 are defined in the fine structures of the NACF. The value of τ_1 is defined as the time duration at the first peak of NACF. The ϕ_1 is the magnitude of NACF at τ_1.

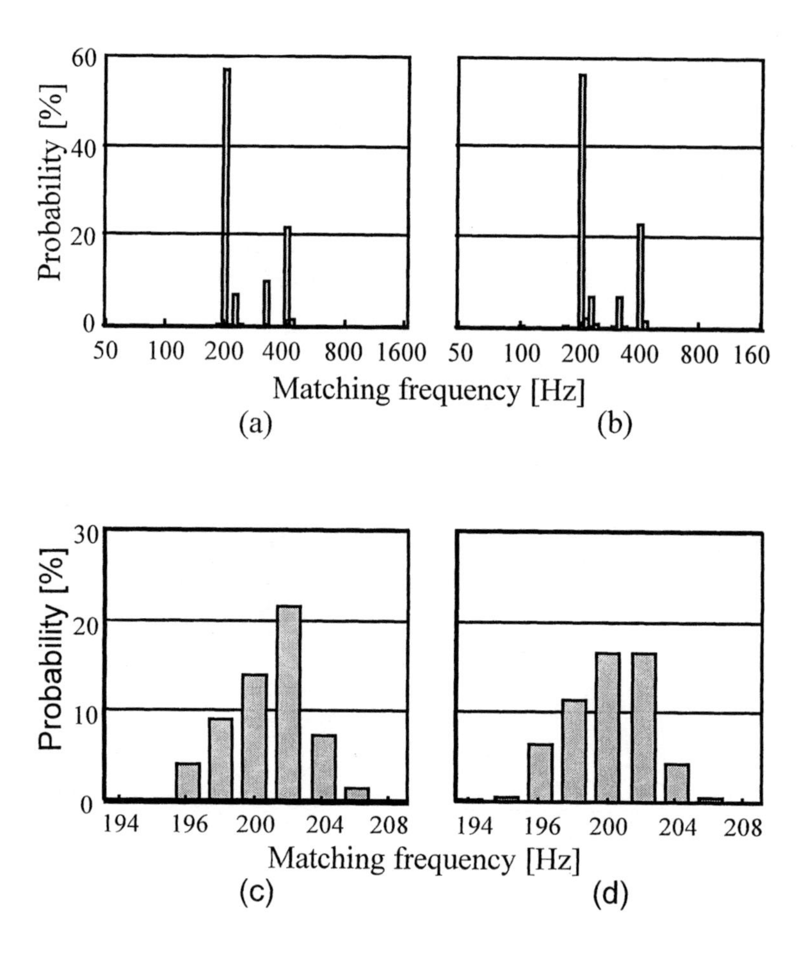

Figure 14. Results of the pitch-matching tests for global data of five subjects: in-phase stimuli (a) and random-phase stimuli (b). Results of the test around 200 Hz (between 183.3 and 218 Hz) of global data from five subjects: in-phase stimuli (c) and random-phase stimuli (d).

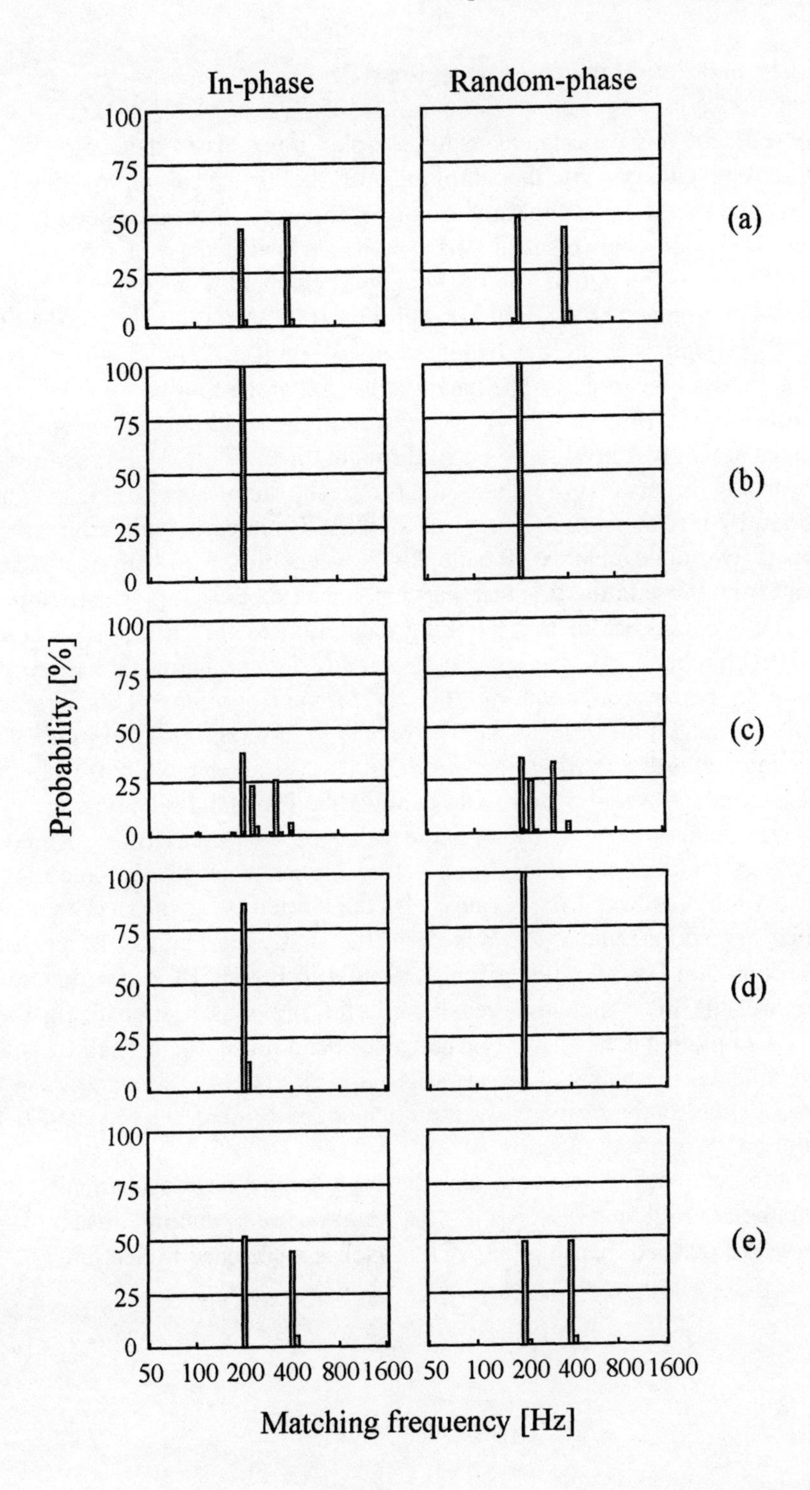

Figure 15. Results of the pitch-matching tests for each subject. Symbols (a) through (e) represent different subjects.

4.3 Pitch perception for complex noises

The purpose of this experiment using complex noises is to determine whether pitch perception is changed by the amplitude of the first peak ϕ_1 of the NACF. The experimental method was the same as that of the experiment described in the previous section. The bandwidth of each partial noise, which consist of the band-pass white noise with a cut-off slope of 1080 dB/oct., was changed. The center frequencies of the bandnoise components were 600 Hz, 800 Hz, 1000 Hz, 1200 Hz and 1400 Hz. Here, the complex signals including bandpass noises with different center frequencies are called here "complex noise". The bandwidths (Δf) of the four components were 40, 80, 120 and 160 Hz (Figure 16(a)). Their waveforms without any specific periodical envelopes are shown in Figures 16(b) through 16(e). The NACFs for four conditions are shown in Figures 16(b') through 16(e'). The amplitude of the maximum peak (indicated by arrows in the figures) in the NACF increases with decreasing Δf. Five musicians (two male and three female, 20–25 years old), who with one exception were different from those in the first test, participated as subjects in this experiment.

The probabilities of the matching data counted for each 1/12-octave band are shown in Figures 17(a) through 17(d). All histograms show that there is a strong tendency to perceive a pitch of 200 Hz for any stimulus. This agrees with the prediction based on the value of τ_1. The results in Figures 18(a) through 18(d) indicate that a stimulus with a narrow bandwidth gives a stronger pitch corresponding to 200 Hz than does a stimulus with a wide bandwidth. The standard deviation (SD) for the perceived pitches increased because the value of ϕ_1 decreased as Δf increased. The probabilities matched to 400 Hz (one octave higher than 200 Hz) keep increasing as the bandwidth narrows. This is caused by the similarity of the octave relation under the pitch perception which also appear in the first experiment. The probability of a pitch around 200 Hz being identified is plotted in Figure 18 as the function of ϕ_1. In this Figure, the pitch-matching result from the previous section using the complex tones is also plotted at $\phi_1 = 1.0$. For narrower-band noise, the probability of a pitch of the fundamental frequency increases as the magnitude of the 5 ms peaks in the NACF increases. Thus, as the ϕ_1 increases the pitch strength increases ($r = 0.98$). This result also supports Wightman's theory.

Individual differences can also be seen in the results obtained in tests with complex noises (Sakai *et al.* [33]). So far we have found that the missing fundamental can be well described by using the ACF model as mentioned in Section 2.

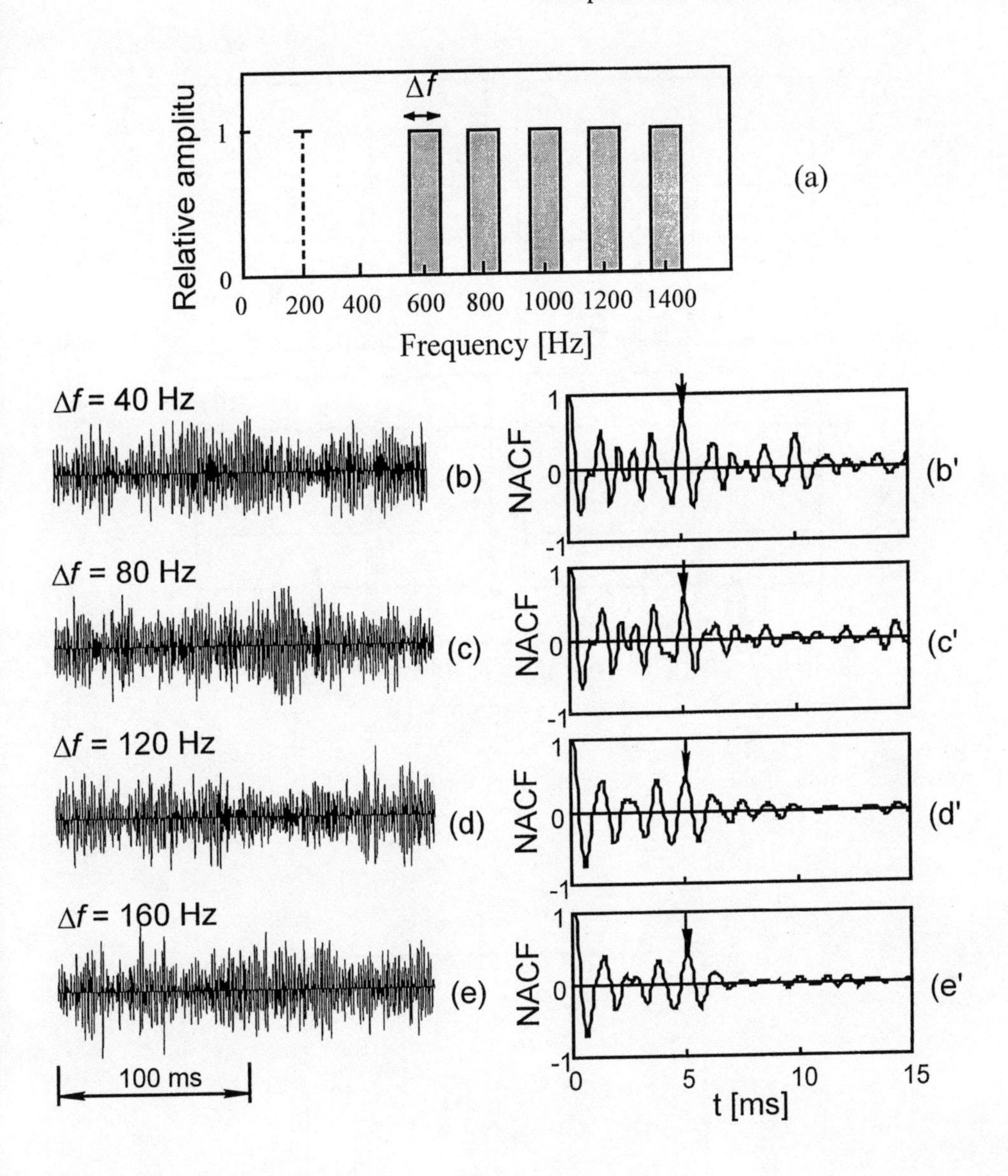

Figure 16. "Complex noise" containing the center frequencies: 600 Hz, 800 Hz, 1000 Hz, 1200 Hz and 1400 Hz used (a). Its fundamental frequency is around 200 Hz. The Δf represents the bandwidth. Waveforms of the four complex noises applied: Δf = 40 Hz (b); Δf = 80 Hz (c); Δf = 120 Hz (d); and Δf = 160 Hz (e). The NACFs of the stimuli: Δf = 40 Hz (b'); Δf = 80 Hz (c'); Δf = 120 Hz (d'); and Δf = 160 Hz (e').

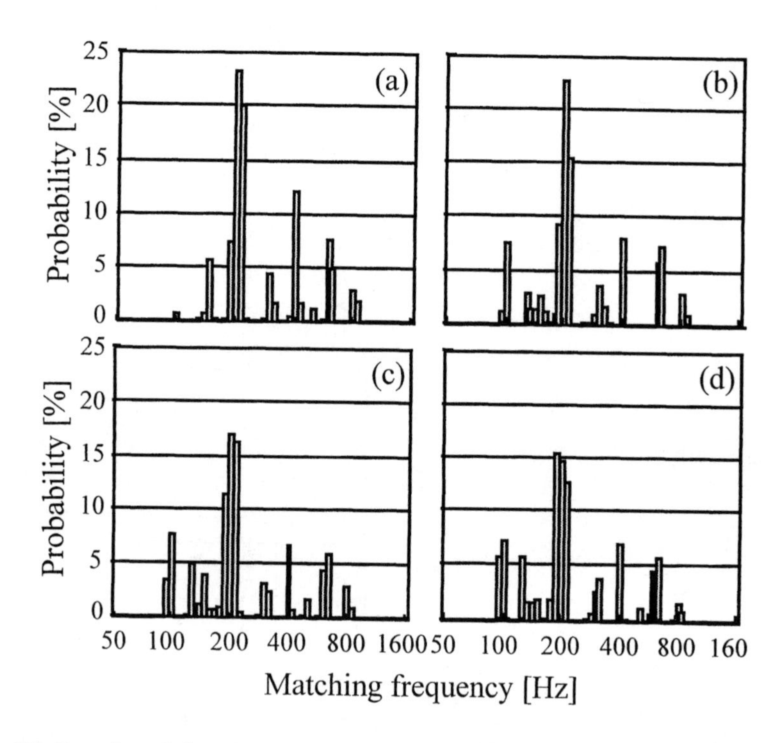

Figure 17. Results of the pitch-matching tests for the global data of five subjects: Δf = 40 Hz (a); Δf = 80 Hz (b); Δf = 120 Hz (c); and Δf = 160 Hz (d).

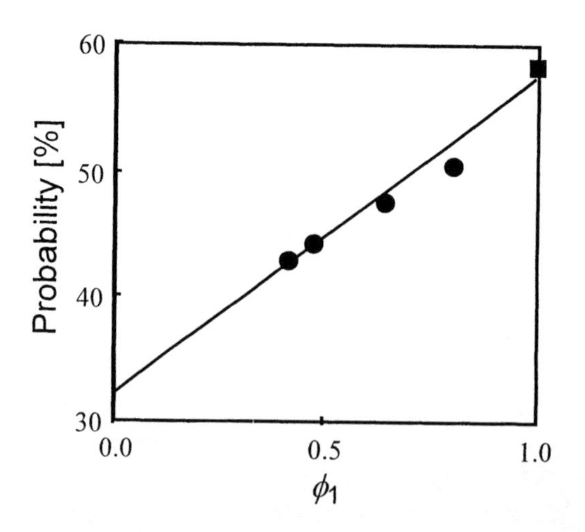

Figure 18. Relationship between ϕ_1 and probability of the pitch being within 200±16 Hz ($r = 0.98$, $p < 0.01$). For reference, the plot (■) at $\phi_1 = 1$ is the result of the first tests.

4.4 Loudness

Loudness judgments were performed while changing the ACF of bandpass noise within the critical band under fixed conditions of the sound pressure level (SPL = 74 dB) and of other temporal and spatial factors described in Section 2 (Merthayasa and Ando [35]). The effective duration of ACF, τ_e, or the repetitive feature, of the bandpass noise of 1 kHz center frequency was controlled by the bandpass filter slope (48, 140 and 1080 dB/oct., which is mixed by the combination of the former two digital filters). The duration of the stimuli was 3 s, the rise and fall times were 250 ms, and the interval between stimuli was 1 s. In this test, the bandwidth Δf of stimuli defined by the –3 dB attenuation of low and high cut-off frequency was kept constant. In fact, Δf was set at "0 Hz" with only the slope components controlling the wide range of τ_e (= 3.5–52.6 ms). The paired-comparison method for judgments was used by more than six students of normal hearing ability. Since the subjects sat in an anechoic chamber facing the loudspeaker 0.90±0.01 m away, the IACC was kept constant at nearly unity.

As shown in Figure 19, the loudness of the bandpass noise within the critical band is not constant. Scale values of loudness indicate a minimum at a certain bandwidth, when the filter slope of 1080 dB/oct. is used. This result differs from the results reported by Zwicker *et al.* [36].

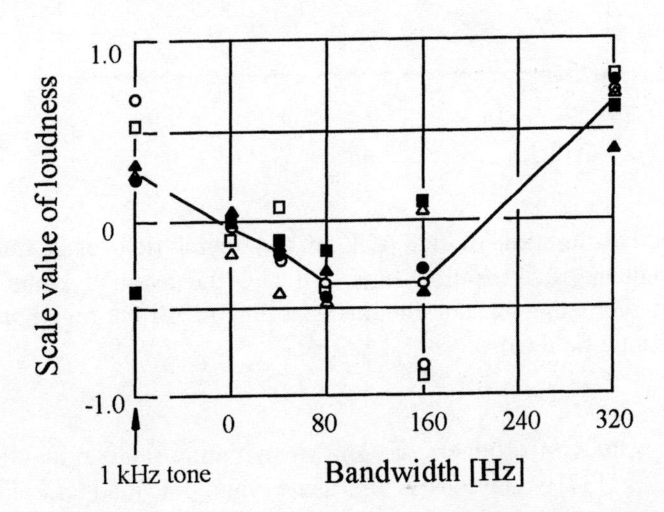

Figure 19. Scale values of the loudness of bandpass noises as a function of its bandwidth centered on 1 kHz. The cutoff slope of the filter used was 1080 dB/oct. Different symbols indicate the scale values obtained with different subjects (six subjects).

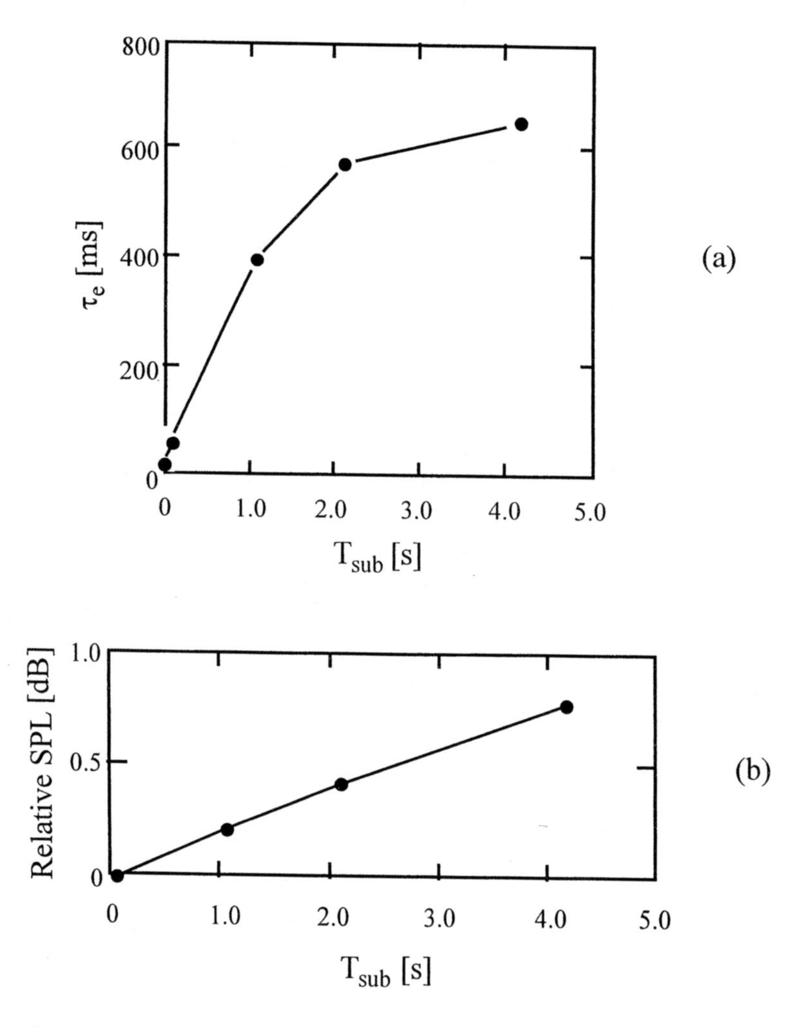

Figure 20. Effective duration of the ACF of the sound field as a function of the subsequent reverberation time (a); and the average value of loudness obtained by the constant method as a function of the reverberation time of the sound field (b).

The scale values of loudness are shown in Figure 4 as a function of log τ_e (Merthayasa *et al.* [37]). Obviously, the scale value of loudness increases with increasing value of τ_e. The statistical analysis for scale values with different values of τ_e indicates significant difference ($p < 0.01$) except for the stimuli of the pure tone and stimuli with log $\tau_e = 1.72$ ($\tau_e = 52.6$ ms). It can be demonstrated that the degree of repetitien of stimuli contributes to the loudness. Since there is no significant difference in loudness between the pure tone and the "0 Hz" bandwidth signal produced by the use of a filter with a slope of 1080 dB/oct., it is recommended to use a sharp slope-filter in the hearing experiment, which corresponds to the sharpness of the filter in the auditory system (Katsuki *et al.* [10]). A similar tendency is observed in that, under the condition of fixed SPL, the value of τ_e increases as the reverberation increases. Accordingly, loudness increases as the reverberation time increases, as shown in Figure 20 (Ono and Ando [38]).

It is worth noting that the loudness does not depend on the IACC under conditions in which the sound pressure level at both ear entrances is fixed (Merthayasa *et al.* [39]). This confirms the results obtained using headphone reproduction (Chernyak and Dubrovsky [40]; Dubrovskii and Chernyak [41]).

5 Calculations of speech intelligibility of each single syllable in relation to the four orthogonal factors extracted from the ACF of both source and sound-field signals

This section discusses the way that the speech intelligibility (SI) of each syllable can be described in terms of the distance between the template source signal and a sound-field signal. In the calculation of this distance, only four independent factors extracted from the ACF of the direct sound as a template, and signals of sound fields are utilized (Section 2.2).

Concerning global speech intelligibility of sound fields, a speech transmission index (STI) has been proposed (Houtgast *et al.* [42]). A distance between the template source signal and sound-field signal is introduced by applying the four orthogonal factors described in Section 2.

Let S_K^T be the isolated template-syllable K, and let S_X^{SF} be another syllable X in the sound field, where T signifying the template and SF being the sound field. The distance between S_K^T and S_X^{SF} for one of four factors is given by

$$d(x) = D_x\left(S_X^{SF}, S_K^T\right) = \left|C_X^{SF}, C_K^T\right| \tag{27}$$

where $x = \tau_e$, τ_1, ϕ_1 or $\Phi(0)$, which are defined in Figure 2 and the factor $\Phi(0)$ is the energy which is represented at the origin of the delay of ACF; C_K^T and C_X^{SF} are characteristics of an isolated template-syllable S_K^T and that of anothoer syllable of the sound-field S_X^{SF} in the brain, respectively; and where

$$D_{\tau_e}\left(S_X^{SF}, S_K^T\right) = \left\{\sum_{i=1}^{n}\left|\log\left(\tau_e^i\left(S_X^{SF}\right)\right) - \log\left(\tau_e^i\left(S_K^T\right)\right)\right|\right\} \Big/ n$$

$$D_{\tau_1}\left(S_X^{SF}, S_K^T\right) = \left\{\sum_{i=1}^{n}\left|\log\left(\tau_1^i\left(S_X^{SF}\right)\right) - \log\left(\tau_1^i\left(S_K^T\right)\right)\right|\right\} \Big/ n \tag{28}$$

$$D_{\phi_1}\left(S_X^{SF}, S_K^{T}\right) = \left\{ \sum_{i=1}^{n} \left| \log\left|\phi_1^i\left(S_X^{SF}\right)\right| - \log\left|\phi_1^i\left(S_K^{T}\right)\right| \right| \right\} \Big/ n$$

$$D_{\Phi(0)}\left(S_X^{SF}, S_K^{T}\right) = \left\{ \sum_{i=1}^{n} \left| \log\left(\Phi(0)^i\left(S_X^{SF}\right) / \Phi(0)^{\max}\left(S_K^{T}\right)\right) - \log\left(\Phi(0)^i\left(S_X^{SF}\right) / \Phi(0)^{\max}\left(S_K^{T}\right)\right) \right| \right\} \Big/ n$$

where n is the frame number of the running ACF. Let the number of syllables of the same category with S_X^{SF} be N. Japanese syllables can be categorized by the type of consonant and vowel as indicated in Table 3 because listeners does not confuse syllables over the catogory [43]. The intelligibility of the syllable K for one of the four factors may be obtained as

$$\Psi_k(x) = 100N \exp\left(-\frac{d_k(x)}{d_1(x)} \cdots \frac{d_k(x)}{d_{k-1}(x)} \frac{d_k(x)}{d_{k+1}(x)} \cdots \frac{d_k(x)}{d_N(x)} \right) \tag{29}$$

Table 3. Categorization of each Japanese syllable.

(a) Unvoiced consonant

	Vowel	Consonant K	S	T	H	P
Not contracted (Category A)	A	KA	SA	TA	HA	PA
	I	KI	SI	TI	HI	PI
	U	KU	SU	TU	HU	PU
	E	KE	SE	TE	HE	PE
	O	KO	SO	TO	HO	PO
Contracted (Category B)	YA	KYA	SYA	TYA	HYA	PYA
	YU	KYU	SHU	TYU	HYU	PYU
	YO	KYO	SHO	TYO	HYO	PYO

(b) Voiced consonant

	Vowel	Consonant N	M	Y	R	W	G	Z	D	B
Not contracted (Category C)	A	NA	MA	YA	RA	WA	GA	ZA	DA	BA
	I	NI	MI	-	RI	-	GI	ZI	-	BI
	U	NU	MU	YU	RU	-	GU	ZU	-	BU
	E	NE	ME	-	RE	-	GE	ZE	DE	BE
	O	NO	MO	YO	RO	-	GO	ZO	DO	BO
Contracted (Category D)	YA	NYA	MYA	-	RYA	-	GYA	ZYA	-	BYA
	YU	NYU	MYU	-	RYU	-	GYU	ZYU	-	BYU
	YO	NYU	MYO	-	RYU	-	GYO	ZYO	-	BYO

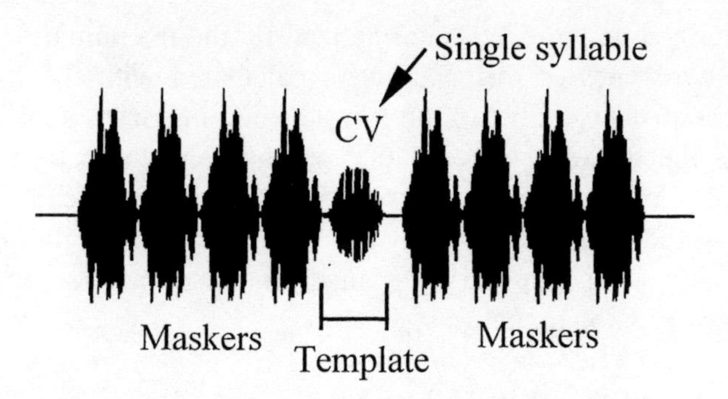

Figure 21. Example of a waveform of Japanese single syllables between artificial non-meaning forward and backward maskers. The direct sound without maskers used as a template.

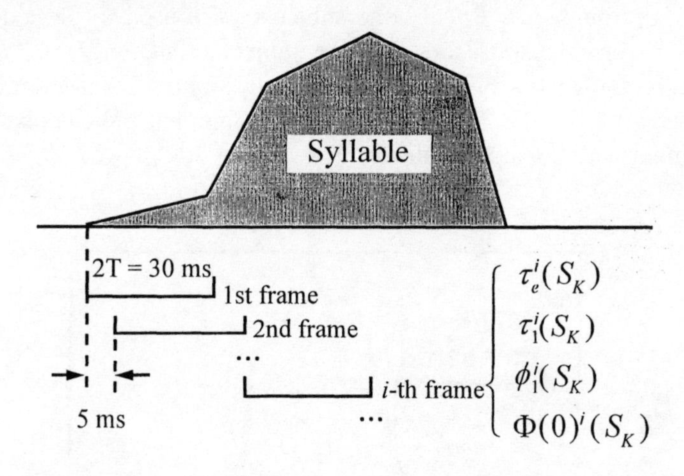

Figure 22. Running ACF of the syllable. Four orthogonal factors are obtained from each frame of the running ACF, integration interval 2T = 30 ms, with a running step of 5 ms for both the template-syllable and the syllable in sound field.

Let us now demonstrate an example of studying the intelligibility of the sound field which consists of the direct sound and the single reflection. The amplitude of reflection was the same as that of the direct sound, and the delay time of the reflection Δt_1 was varied in the range between 0 and 480 ms. The test signals consisted of 50 Japanese monosyllables with maskers (Figure 21). The direct sound without maskers was used as a template. The running ACFs, integration interval 2T = 30 ms, with the running step of 5 ms were calculated (Figure 22). Four orthogonal factors were obtained from each frame of running ACF. The important former half parts of the ACFs $\phi(0) < 0.5$ (Figure 23) of both a template and a test syllable were analyzed. The value of $\phi(0)$ is defined as the energy normalized by the maximum value among all frames. Actually, two types of distances were calculated with the four orthogonal factors. One was the distance between the template and the direct sound with makers, and the other was the distance between the template and the reflected sound with maskers. The shorter of those distances was selected in the calculation of syllable intelligibility. Using the distances given by eqn (28), the intelligibility of all syllables can be calculated by using eqn (29). Distances for four factors can be combined linearly due to the expression given by

$$\Psi_k = a\Psi_k(\tau_e) + b\Psi_k(\tau_1) + c\Psi_k(\phi_1) + d\Psi_k(\Phi(0)) \tag{30}$$

where a, b, c and d are the coefficients to be evaluated. Results of both calculated and tested intelligibility for single syllables belonging to each category are demonstrated in Figures 24(a) through (d). Twenty-one subjects were used in the experiments, so that about a 5% error results for a single subject's judgement in these figures. Averaged results for each category are shown in Figure 25. Mean values of all syllables are shown in Figure 26. It is quite remarkable that the calculated values are in good agreement with the tested values.

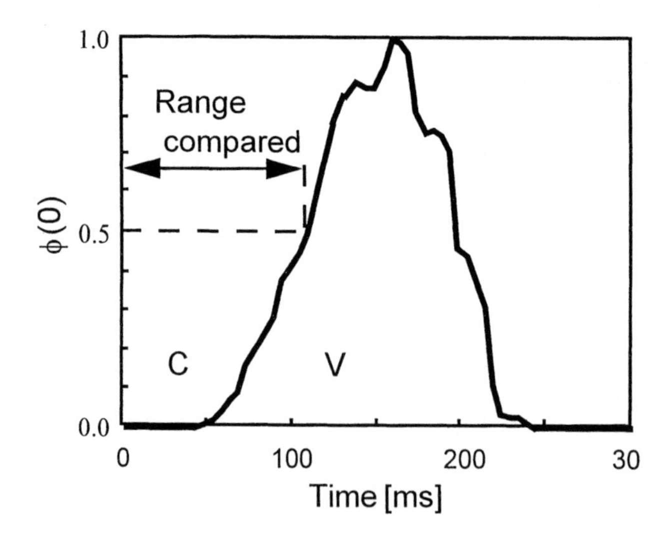

Figure 23. The power function of a syllable. The important former half parts of running ACFs $\phi(0) < 0.5$ of both a template and a test syllable were analyzed.

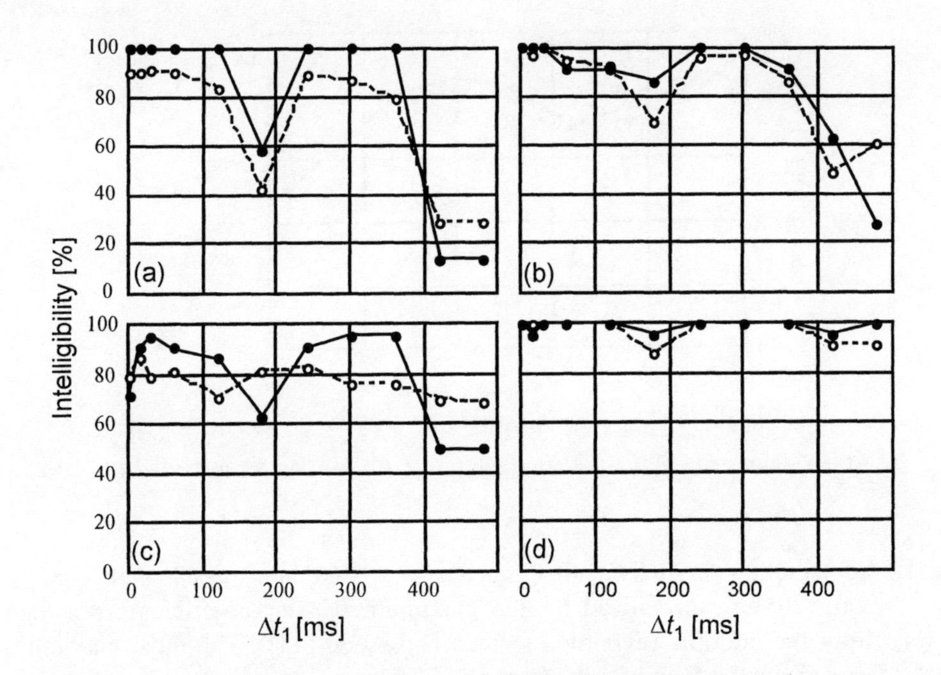

Figure 24. Results of calculated and tested intelligibilities for single syllables belonging to each category: /ha/ (a); /be/ (b); /hya/ (c); /nyu/ (d). Twenty-one subjects were used in the experiments

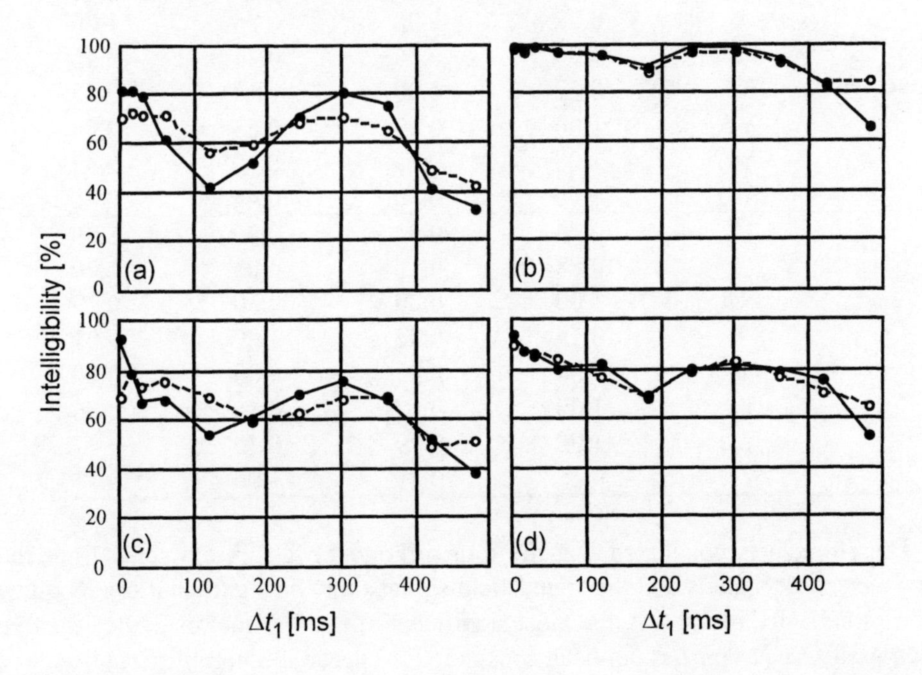

Figure 25. Averaged results of calculated and tested intelligibilities for each category: category A (a); category B (b); category C (c); category D (d).

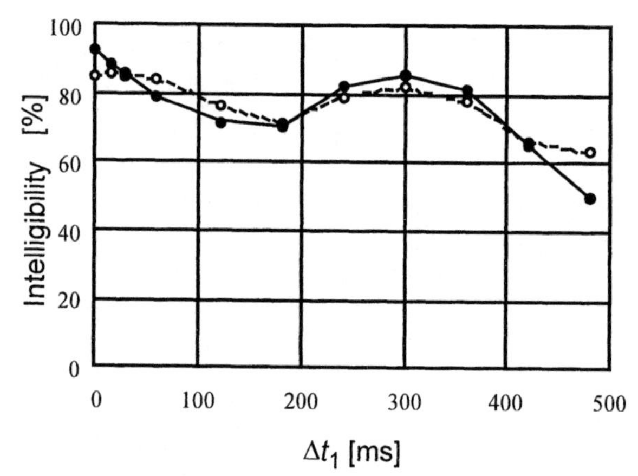

Figure 26. Mean values of calculated and tested intelligibilities for all syllables.

Table 4. Contribution of each factor extracted from ACF for speech intelligibility. Values were normalized by the maximum of four coefficients obtained from the multiple regression (Shoda and Ando [44]). The most maximum values may be found for the factor τ_e.

Consonant	Vowel	τ_e	τ_1	ϕ_1	$\Phi(0)$
Unvoiced	A	0.50	0.60	0.36	1.00
	U	0.07	0.08	0.25	1.00
	E	1.00	0.66	0.23	0.46
	O	0.62	1.00	0.13	0.25
	YA	1.00	0.17	0.55	0.56
	YU	1.00	0.19	0.30	0.47
	YO	1.00	0.30	0.41	0.88
Voiced	A	0.74	0.92	0.89	1.00
	I	0.88	0.38	0.01	1.00
	U	1.00	0.80	0.30	0.29
	E	0.19	0.42	1.00	0.01
	O	1.00	0.25	0.09	0.80
	YA	1.00	0.17	0.55	0.56
	YU	1.00	0.19	0.30	0.47
	YO	0.25	0.21	1.00	0.51

Therefore, it is concluded that the four orthogonal factors extracted from the ACF of the source signals and the sound-field signals may be significant to recognize speech. As indicated in Table 4, the most significant of the four factors is the effective duration of the ACF. This has not been considered in speech intelligibility studies.

References

[1] Ando, Y., *Concert Hall Acoustics*, Springer-Verlag, Heidelberg, 1985.

[2] Ando, Y., Yamamoto, K., Nagamatsu, H., and Kang, S. H., Auditory brainstem response (ABR) in relation to the horizontal angle of sound incidence, *Acoust. Lett.*, **15**, 57-64, 1991.

[3] Ando, Y., Kang, S. H., and Nagamatsu, H., On the auditory-evoked potential in relation to the IACC of sound field, *J. Acoust. Soc. Jpn.* (*E*), **8**, 183-190, 1987.

[4] Ando, Y., Kang, S. H., and Morita, K., On the relationship between auditory-evoked potential and subjective preference for sound field, *J. Acoust. Soc. Jpn.* (*E*), **8**, 197-204, 1987.

[5] Ando, Y., Evoked potentials relating to the subjective preference of sound fields, *Acustica*, **76**, 292-296, 1992.

[6] Ando, Y., and Chen, C., On the analysis of autocorrelation function of α-waves on the left and right cerebral hemispheres in relation to the delay time of single sound reflection, *J. Arch. Plan. Environ. Eng.*, Architectural Institute of Japan (AIJ), **488**, 67-73, 1996.

[7] Chen, C., and Ando, Y., On the relationship between the autocorrelation function of the α-waves on the left and right cerebral hemisphere and subjective preference for the reverberation time of music sound field, *J. Arch. Plan. Environ. Eng.*, Architectural Institute of Japan (AIJ), **489**, 73-80, 1996.

[8] Nishio, K., and Ando, Y., On the relationship between the autocorrelation function of the continuous brain waves and the subjective preference of sound field in change of the IACC, *J. Acoust. Soc. Am.*, **100** (A), 2787, 1996.

[9] Ando, Y., *Architectural Acoustics, Blending Sound Sources, Sound Fields, and Listeners*, AIP Press/Springer-Verlag, New York, 1998.

[10] Katsuki, Y., Sumi, T., Uchiyama, H., and Watanabe, T., Electric responses of auditory neurons in cat to sound stimulation, *J. Neurophysiol.* **21**, 569-588, 1958.

[11] Kiang, N. Y. –S., *Discharge Pattern of Single Fibers in the Cat's Auditory Nerve*, MIT Press, Cambridge, MA, 1965.

[12] Roederer, J.G., *The Physics and Psychophysics of Music, an Introduction,* 3rd edn., Springer-Verlag, New York, 1995

[13] Wightman, F. L., Pitch and stimulus fine structure, *J. Acoust. Soc. Am.* **54**, 397-406, 1973.

[14] Ando, Y., and Alrutz, H., Perception of coloration in sound fields in relation to the autocorrelation function, *J. Acoust. Soc. Am.*, **71**, 616-618, 1982.

[15] Damaske, P., and Ando, Y., Interaural crosscorrelation for multichannel loudspeaker reproduction, *Acustica*, **27**, 232-238, 1972.

[16] Nakajima, T., and Ando, Y., Effects of a single reflection with varied horizontal angle and time delay on speech intelligibility, *J. Acoust. Soc. Am.*, **90**, 3173-3179, 1991.

[17] Ando, Y., and Kurihara, Y., Nonlinear response in evaluating the subjective diffuseness of sound field, *J. Acoust, Soc. Am.*, **80**, 833-836, 1986.

[18] Singh, P. K., Ando, Y., and Kurihara, Y., Individual subjective diffuseness responses of filtered noise sound fields, *Acustica*, **80**, 471-477, 1994.
[19] Sato, S., and Ando, Y., Effects of interaural crosscorrelation function on subjective attributes, *J. Acoust. Soc. Am.,* **100** (A), 2592, 1996.
[20] Keet, M. V., The influence of early lateral reflections on the spatial impression, *Proc. 6th Intern. Congr. Acoust.*, Tokyo, Paper E-2-4, 1968.
[21] Licklinder, J. C. R., A duplex theory of pitch perception, *Experimenta*, **7**, 128-134, 1951.
[22] Meddis, R., and Hewitt, M. J., Virtual pitch and phase sensitivity of a computer model of the auditory periphery. I: Phase identification, *J. Acoust. Soc. Am.* **89**, 2866-2882, 1991.
[23] Meddis, R., and Hewitt, M. J., Virtual pitch and phase sensitivity of a computer model of the auditory periphery. II: Phase sensitivity, *J. Acoust. Soc. Am.* **89**, 2883-2894, 1991.
[24] Wightman, F. L., The pattern-transformation model, *J. Acoust. Soc. Am.* **54**, 407-416, 1973.
[25] Goldstein, J. L., An optimum processor theory for the central formation of the pitch of complex tones, *J. Acoust. Soc. Am.* **54**, 1496-1516, 1973.
[26] Terhardt, E., Pitch, consonance, and harmony, *J. Acoust. Soc. Am.* **55**, 1061-1069, 1974.
[27] Yost, W. A., Hill, R., and Perez-Falcon., T., Pitch and pitch discrimination of broadband signals with rippled power spectra, *J. Acoust. Soc. Am.* **63**, 1166-1173, 1978.
[28] Yost, W. A., A time domain description for the pitch strength of iterated rippled Noise, *J. Acoust. Soc. Am.* **99**, 1066-1078, 1996.
[29] Yost, W. A., Pitch of iterated rippled noise, *J. Acoust. Soc. Am.* **100**, 511-518, 1996.
[30] de Cheveigne, A., Cancellation model of pitch perception, *J. Acoust. Soc. Am.* **103**, 1261-1271, 1998.
[31] Ohgushi, K., On the role of spatial and temporal cues in the perception of the pitch of complex tones, *J. Acoust. Soc. Am.* **64**, 764-771, 1978.
[32] Sumioka, T., and Ando, Y., On the pitch identification of the complex tone by the autocorrelation function (ACF) model, *J. Acoust. Soc. Am.*, **100** (A), 2720, 1996.
[33] Sakai, H., Sumioka, T., and Ando, Y. (unpublished). Pitch identification using the autocorrelation function model.
[34] Lundeen, C., and Small, A. M., Jr., The influence of temporal cue on the strength of periodicity pitches, *J. Acoust. Soc. Am.* **75**, 1578-1587, 1984.
[35] Merthayasa, I Gde N., and Ando, Y., Variation in the autocorrelation function of narrow band noises: their effect on loudness judgment, *Japan and Sweden Symposium on Medical Effects of Noise*, 1996.
[36] Zwicker, E., Flottorp, G., and Stevens, S.S., Critical band width in loudness summation, *J. Acoust. Soc. Am.,* **29**, 548-557, 1957.
[37] Merthayasa, I Gde N., Hemmi, H., and Ando, Y., Loudness of a 1 kHz pure tone and sharply (1080 dB/oct.) filtered noises centered on its frequency, *Mem. Grad.*

School Sci. & Technol., Kobe Univ., **12-A**, 147-156, 1994.
[38] Ono, K., and Ando, Y., A study on loudness of sound field in relation to the reverberation time, *Reports of Architectural Institute of Japan*, Kinki Chapter, 121-124 (in Japanese), 1996.
[39] Merthayasa, I Gde N., Ando, Y., and Nagatani, K., The effect of interaural cross correlation (IACC) on loudness in a free field, *Proc. Seminar National & Pameran Akustik '95* (Indonesia), 174-184, 1995.
[40] Chernyak, R. I., and Dubrovsky, N. A., Pattern of the noise images and the binaural summation of loudness for the different interaural correlation of noise, *Proc. 6th Intern. Congr. Acoust.*, Tokyo, Paper A-3-12, 1968.
[41] Dubrovskii, N. A., and Chernyak, R. I., Binaural loudness summation under varying degrees of noise correlation, *Sov. Phys. Acoust.*, **14**, 326-332, 1969.
[42] Houtgast, T., Steeneken, H. J. M., and Plomp, R., Predicting speech intelligibility in rooms from the modulation transfer function. I. General room acoustics, *Acustica*, **46**, 60-72, 1980.
[43] Korenaga, Y., and Ando, Y., A method of calculating intelligibility of sound field in relation to temporal structure of reflections - On the trend of syllable confusion under sound fields composed of a direct sound and up to two reflections, *J. Acoust. Soc. Jpn.* **52**, 940-947 (in Japanese), 1996.
[44] Shoda, T., and Ando, Y., Calculation of speech intelligibility using four orthogonal factors extracted from the autocorrelation function of source and sound field signals, *Proc. 16th Intern. Congr. Acoust.*, Seattle, 2163-2164, 1998.
[45] Ando, Y., Subjective preference in relation to objective parameters of music sound fields with a single echo, *J. Acoust. Soc. Am.*, **62**, 1436-1441, 1977.
[46] Kuttruff, H., *Room Acoustics*. 3rd edn., Elsevier Applied Science, London, 1991.
[47] Sabine, W.C., *Reverberation. The American Architect and the Engineering Record,* Sabine, W. C. Prefaced by Beranek, L.L. (1992). Collected papers on acoustics, Peninsula Publishing, Los Altos, CA, Chapter 1, 1900.

Appendix: Orthogonal factors of sound fields

In order to discuss fundamental subjective attributes of sound fields, let us consider the transmission of sound from a source point to binaural entrances and extract factors of the sound field. Let $p(t)$ be the source signal as a function of time t, and let $g_l(t)$ and $g_r(t)$ be impulse responses between the source point r_0 and the binaural entrances. Then the sound signals arriving at the entrances are expressed, according to Ando [45], by

$$f_l(t) = p'(t) * g_l(t)$$

and (31)

$$f_r(t) = p'(t) * g_r(t)$$

where the asterisk denotes convolution.

The impulse responses $g_{l,r}(t)$ include the direct sound and reflections $w_n(t - \Delta t_n)$ in the room as well as the head related impulse responses $h_{nl,r}(t)$:

$$g_{l,r}(t) = \sum_{n=0}^{\infty} A_n w_n(t - \Delta t_n) * h_{nl,r}(t) \tag{32}$$

where n denotes the number of reflections with horizontal angle ξ_n and elevation η_n, $n = 0$ signifies the direct sound ($\xi_0 = 0$, $\eta_0 = 0$); $A_0 w_0(t - \Delta t_0) = \delta(t)$, $\Delta t_0 = 0$, $A_0 = 1$, $\delta(t)$ being the Dirac delta function; A_n is the pressure amplitude of the nth reflection $n > 0$; w_n(t) is the impulse response of the walls for each path of reflection arriving at the listener; Δt_n is the delay time of reflection relative to that of the direct sound; and $h_{nl,r}(t)$ are impulse responses for diffraction of the head and pinnae for the single sound direction of n. Eqn (31) then becomes

$$f_{l,r}(t) = \sum_{n=0}^{\infty} p'(t) * A_n w_n(t - \Delta t_n) * h_{nl,r}(t) \tag{33}$$

If the source has a certain directivity, $p(t)$ is replaced by $p_n(t)$.

Orthogonal physical factors of sound fields may be extracted from eqn (33) (Ando, [9]), such as:

1. The initial time delay gap between the direct sound and the first reflection:

$$\Delta t_1 = (d_1 - d_0)/c \tag{34}$$

Note that the amplitude A_1 (= d_0 / d_1) is not physically independent; for example, the value of A_1 is closely related to Δt_1 in such a way that

$$\Delta t_1 = d_0(1/A_1 - 1)/c \tag{35}$$

In addition, the initial time delay gap between the direct sound and the first reflection, Δt_1, is statistically related to Δt_2, Δt_3, ..., which depend on the dimensions of the room. In fact, the echo density is proportional to the square of the time delay (Kuttruff [46]). Thus, the initial time delay gap Δt_1 is regarded as a representation of both sets of Δt_n and A_n ($n = 1, 2, ...$).

2. Another parameter is the set of the impulse responses of the nth reflection, $w_n(t)$, that is given by Sabine's formula [47]; the subsequent reverberation time is approximately calculated by

$$T_{\text{sub}} \approx \frac{KV}{\overline{\alpha}S} \tag{36}$$

where K is a constant (about 0.162), V is the volume of the room, S is the total surface and $\overline{\alpha}$ is the average absorption coefficient of walls, and $\overline{\alpha}S$ is given by the summation of the absorption of each surface.

3. Two sets of the head-related impulse responses for the two ear is, $h_{nl,r}(t)$, constitute the remaining objective parameter. To represent the interdependence between two impulse responses, we can introduce a single factor: the inter-aural cross-correlation function between the sound signals at the two ears, $f_l(t)$ and $f_r(t)$. This function is defined by

$$\Phi_{lr}(t) = \lim_{T\to\infty} \frac{1}{2T} \int_{-T}^{+T} f_l'(t) f_r'(t+\tau)\mathrm{d}t, \quad |\tau| \leq 1 \text{ ms} \tag{37}$$

where $f_l'(t)$ and $f_r'(t)$ are approximately obtained by signals $f_{l,r}(t)$ after passing through the A-weighted network, which corresponds to the ear sensitivity $s(t)$. The external and the middle ear may characterize the ear sensitivity.

The normalized inter-aural cross-correlation function is defined by

$$\phi_{lr}(\tau) = \frac{\Phi_{lr}(\tau)}{\sqrt{\Phi_{ll}(0)\Phi_{rr}(0)}} \tag{38}$$

where $\Phi_{ll}(0)$ and $\Phi_{rr}(0)$ are respectively the autocorrelation functions at $\tau = 0$ for the left and right ears, and they are the sound energies arriving at both ears.

From the inter-aural cross-correlation function, the following four independent factors are extracted:

1. The magnitude of the inter-aural cross-correlation is defined by

$$\text{IACC} = |\phi_{lr}(\tau)|_{max} \tag{39}$$

for the possible maximum inter-aural time delay, say, $|\tau| \leq 1$ ms.

2. The inter-aural delay time at which IACC is defined as shown is the τ_{IACC}.

3. The width of the inter-aural cross-correlation function defined by the interval of delay time at a value of δ below the IACC, that may correspond to the JND of the IACC, is given by the W_{IACC}.

4. The denominator of eqn (38),

$$\text{LL} = \sqrt{\Phi_{ll}(0)\Phi_{rr}(0)} \tag{40}$$

is the geometrical mean of the sound energies arriving at the two ears.

Chapter 5

The sound field for listeners in concert halls and auditoria

J.S. Bradley
Institute for Research in Construction, National Research Council, Montreal Road, Ottawa, Canada, K1A 0R6

Abstract

This chapter reviews our current understanding of the importance of sound levels and spatial impression on the acoustical requirements for listeners in concert halls and auditoria. The accuracy of new techniques for predicting early, late and total sound levels are first demonstrated. The larger prediction errors at lower frequencies are attributed to the excess attenuation of the seat-dip effect, and new results indicate that the magnitude of this attenuation is related to properties of the hall ceiling. Subjective evaluations of the perceived strength of bass and treble sounds are related to particular level measures. New work on spatial impression shows that it has two subjective dimensions: apparent source width ASW and listener envelopment LEV. Perceived ASW is related to the strength of early arriving lateral reflections and LEV is related to the strength of later arriving lateral reflections. This more complete understanding of spatial impression is used to explain a number of practical problems in concert halls.

1 Introduction

For about 100 years the sound field for listeners in concert halls and auditoria has been assessed in terms of reverberation time [1]. More recent work has shown that subjective impressions of reverberation are more closely related to the early decay time. Other measures such as early-to-late ratios and the centre time [2–4] have come into use as better indicators of the trade-off between clarity and reverberance. These are all essentially uni-dimensional quantities. They consider only temporal effects.

With the recognition of the importance of spatial impression and spatial effects [5,6], concert hall acoustics became a problem with both temporal and spatial dimensions. Other modern studies suggest that there are four or five important orthogonal characteristics to be considered [7–9]. We are only beginning to understand the details of each of these factors.

This chapter reviews the current state of our understanding of spatial impression and sound levels for listeners in auditoria. Sound levels are perhaps one of the most neglected but more important means of assessing acoustical conditions for an audience. The effect of a hall on sound levels relates directly to the maximum possible loudness and to the dynamic range of sounds that the audience hears. The relative levels of lower and higher frequency sounds influence the tone quality of the sounds we hear. In addition, sound levels are also a key factor to perceived spatial impression. Both the apparent width of the source and the amount of perceived listener envelopment are very much influenced by aspects of the sound level.

2 Sound levels in halls

The effect of an auditorium on the sound levels heard by listeners can be described by the strength or relative sound level G. This is equivalent to the ratio of the sound level measured at some point in the hall to the level of the same source, at a reference distance, in a free field such as in an anechoic test room. Some like to refer to this quantity as the hall 'gain' in analogy to the gain of a simple amplifier. G is defined as follows:

$$G = 10 \log \left\{ \int_0^\infty p^2(t)dt \Big/ \int_0^\infty p_A^2(t)dt \right\}, \ dB \tag{1}$$

where $p(t)$ is the instantaneous pressure in the room impulse response and $p_A(t)$ is the instantaneous pressure response of the same source at a distance of 10 m in a free field.

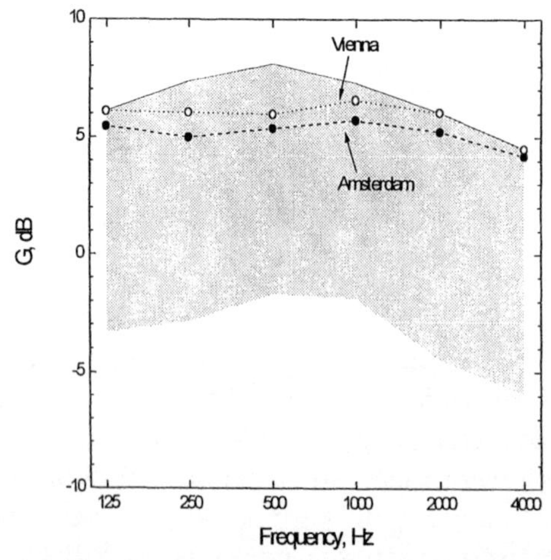

Figure 1. Range of mean G values from the 31 large halls in Table 1 unoccupied. Shown separately are the mean G values for two well known halls: the Vienna Musikvereinssaal and the Amsterdam Concertgebouw.

Table 1. The 31 halls having volumes greater than 10,000 m^3.

Hall and symbol	Volume, m^3	Hall and symbol	Volume, m^3
Mechanics Hall, Worcester, MEC	10,760	Boston Symphony Hall, BOS	18,740
Troy Music Hall, TRY	11,320	Amsterdam Concertgebouw, AMS	18,774
Western Manitoba Centennial Auditorium, Brandon, WMC	12,390	Kennedy Centre Concert Hall, Washington, WAS	19,300
Odense Concert Hall, ODE	14,000	Thomas Hall, Akron, AKR	19,800
Massey Hall, Toronto, MAS	14,190	Opera, National Arts Centre, Ottawa, NAC	20,000
Salle Pleyel, Paris, PLY	15,000	Jubilee Auditorium, Edmonton, JUB	21,480
Musikvereinssaal, Vienna, VNA	15,000	Myerhoff Symphony Hall, Baltimore, BAL	21,500
Festspielhaus, Salzburg, SLZ	15,500	Orpheum Theatre, Vancouver, ORP	22,000
Centre in the Square, Kitchener, KIT	15,600	Manitoba Centennial Auditorium, Winnipeg, MCA	28,571
Academy of Music, Philadelphia, PHI	15,700	Roy Thomson Hall, Toronto, RTH	28,600
Severance Hall, Cleveland, CLV	15,700	Philharmonie Am Gasteig, Munich, MUN	31,000
Orchestra Hall, Detroit, DET	15,700	Hamilton Place, Hamilton, HAM	31,413
Liederhalle,Grosser Sal, Stuttgart, STU	16,000	Salle Wilfrid Pelletier, Montreal. SWP	32,100
Queen Elizabeth Theatre, Vancouver, QET	16,760	Boettcher Hall, Denver, DEN	37,237
Singer Hall, Calgary, CAL	17,000	Tanglewood Music Shed, Lennox, TNG	42,450
Kleinhans Music Hall, Buffalo, BUF	18,220		

G values are typically measured in octave bands and can be recorded for individual measurement positions or averaged to give hall mean values. Figure 1 illustrates the range of hall mean *G* values from 31 large unoccupied auditoria. There is a trend for *G* values to peak at mid-frequencies. Air absorption causes higher frequency *G* values to decrease in all halls and in many halls lower frequency *G* values are also lower than those at mid-frequencies. This figure also identifies the mean *G* values for two well known concert halls, the Amsterdam Concertgebouw and the Vienna Musikvereinssaal.

One can also calculate *G* values separately for the early and late arriving portions of the total sound. *G80,* the relative level of the direct and early arriving sound in the first 80 ms after the arrival of the direct sound, is calculated using eqn (1) but changing the upper limit of the first integral from ∞ to 0.08 s. Similarly the relative level of the late arriving sound, GL, can be calculated using eqn (1) and integrating the room impulse response from 0.08 s to ∞. GL and $G80$ are particularly useful because the early and late portions of the total sound energy tend to be influenced by different physical parameters and can be related to different subjective perceptions. For example, the details of the geometry of the hall can increase the level of early arriving sound and can lead to a sense of increased clarity. Later arriving sound is more influenced by the total amount of sound absorption in the room. When there is increased late arriving sound energy, the sound will be perceived to be more reverberant and enveloping.

2.1 Barron's theory

Barron showed that simple diffuse field theory does not provide accurate predictions of expected sound levels in auditoria [3, 10]. He developed his revised theory that has been found to be a more accurate indication of expected relative sound levels in halls [10]. His theory has the added advantage of being able to predict both early and late relative sound levels as well as the total relative sound level. Of course, the ratio of the early to the late sound energy is the familiar early/late sound ratio, $C80$, which is a measure of clarity.

Barron's theory considers relative sound levels as due to three components: the relative energy of the direct sound d, the early reflected sound e, and the late arriving reflected sound l. These are given by:

$$d = 100/r^2$$

$$e = (31200\ T/V)\ \mathrm{e}^{-0.04\,r/T}\,(1 - \mathrm{e}^{-1.1/T})$$

$$l = (31200\ T/V)\ \mathrm{e}^{-0.04\,r/T}\,(\mathrm{e}^{-1.1/T})$$

where r is the source-to-receiver distance (m), V is the room volume (m^3) and T the reverberation time (s).

The total relative sound level G is simply 10 times the logarithm of the sum of the relative energies of the direct, early and late sound energies,

$$G = 10\log\{d + e + l\},\ \mathrm{dB} \tag{2}$$

Similarly, the relative level of the early sound $G80$ is calculated from the sum of the direct and early reflected sound energy and the late relative level GL from the late arriving sound energy:

$$G80 = 10\log\{d + e\},\ \mathrm{dB} \tag{3}$$

$$GL = 10\log\{l\},\ \mathrm{dB} \tag{4}$$

$C80$, the ratio of early arriving to late arriving sound, is defined as follows:

$$C80 = 10\log\left\{\int_0^{0.08} p^2(t)dt \Big/ \int_{0.08}^{\infty} p^2(t)dt\right\} = G80 - GL,\ dB$$

2.2 Early, late and total levels in auditoria

Barron's theory is a very useful new tool to help in the design and understanding of auditoria. Because the theory tends to successfully predict conditions in well behaved sound fields, deviations of measured values from the predictions of this theory often indicate some unusual feature of a particular hall.

One can assess the validity of the predictions of Barron's theory by comparing predictions with mean measured values from 35 halls. Predicted values were obtained by using the mean reverberation time and the mean source-to-receiver distance in the equations given in the previous section. Figure 2 compares mean measured 1000 Hz *G* values from 35 halls with these predictions. The halls are the 31 larger halls in Table 1 plus four smaller halls with volumes between 3000 and 6000 m^3 to give a wider range of halls. For these mid-frequency total *G* values the predictions tend to be a little higher than measured values, but the average difference is less than 1 dB.

Figure 3 similarly compares average measured 125 Hz total *G* values with predictions. For these lower frequency results, the average difference is 2.4 dB. Similar plots were produced for other octave band frequencies and for the early and late components as well as the total sound levels. From each of these plots the average difference between measured and predicted values was determined. The resulting average differences are plotted in Figure 4. These results indicate that predictions are most accurate at mid-frequencies and that the prediction errors vary depending on whether early, late or total *G* values are considered. If simple diffuse field theory is used to predict these measured *G* values, the prediction errors are approximately 2 dB greater than those in Figure 4.

The larger average prediction errors at lower frequencies in Figure 4 can be attributed to the grazing incidence seat-dip attenuation [11–13]. Sound passing over the audience at near grazing incidence is strongly attenuated at lower frequencies. This affects the early sound more than later arriving sound, which explains why the low-frequency attenuation prediction errors are larger for *G80* than for *GL* values. This excess low-frequency attenuation is discussed further in section 2.4 below.

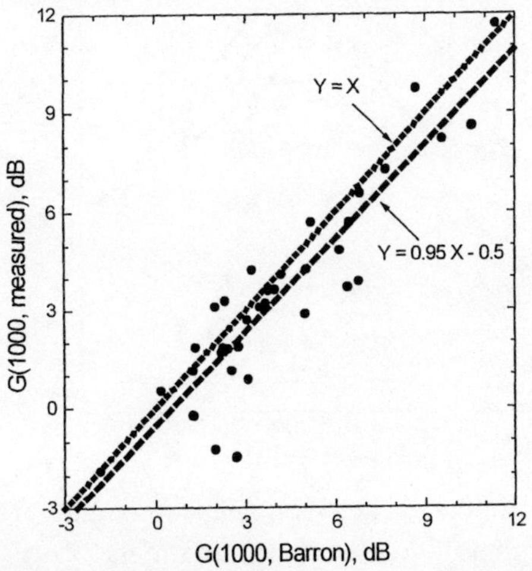

Figure 2. Comparison of measured and predicted average 1000 Hz total *G* values.

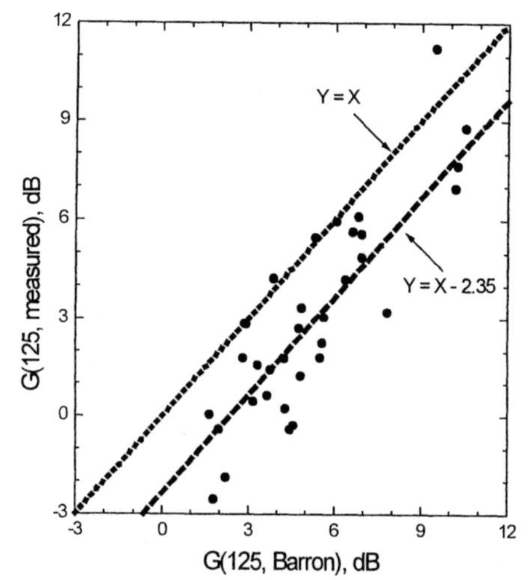

Figure 3. Comparison of average measured and predicted 125 Hz total *G* values.

At 500 and 1000 Hz the Barron theory predictions work best for the early arriving sound (*G80* values). Average measured *G80* values at 500 and 1000 Hz are only a fraction of a decibel less than predicted. However, the prediction errors for the late arriving sound (*GL* values) are a little larger.

The predictions can be improved a little if the early decay time rather than the conventional reverberation time is used in Barron's theory. (The early decay time, EDT, is the reverberation time measured from the slope of the first 10 dB of the decay.) This is justifiable because most of the sound energy is in the initial part of the decay and the early decay time is a more accurate indicator of this energy. For the data from the 35 halls the EDT/RT ratios are approximately 0.9, varying a little with frequency. Figure 5 shows the corresponding set of prediction errors when early, late

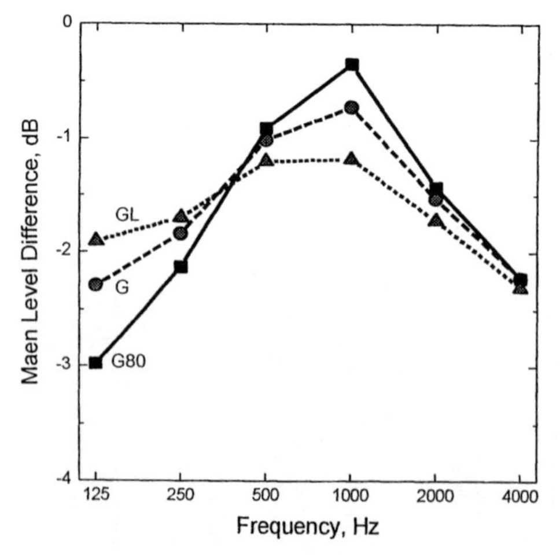

Figure 4. Mean level differences between measured and predicted values using Barron's theory based on reverberation times.

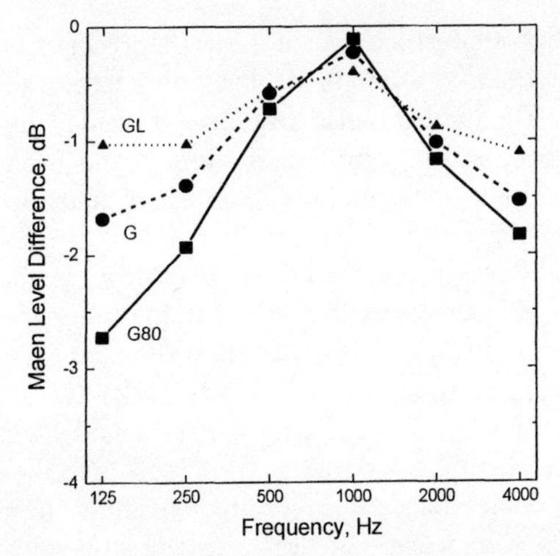

Figure 5. Mean level differences between measured and predicted values using Barron's theory based on early decay times.

and total relative levels are predicted using Barron's theory (eqns (2), (3) and (4)) but replacing reverberation times with early decay times. In all cases prediction errors are reduced compared to the mean errors in Figure 4. At 1000 Hz average predicted levels are very accurate and prediction errors are only a little larger at 500 Hz. At lower frequencies, prediction errors are reduced most for late relative sound levels and the prediction of the early arriving low-frequency *G80* values is only slightly improved because these differences are mostly related to the seat dip attenuation. Predictions are still a little less accurate at higher frequencies, and it has not been established why this is so. It is possible that high-frequency sound levels in the halls are on average reduced due to scattering losses.

2.3 Subjective ratings of bass and treble levels

It would seem likely that the relative levels of lower and higher frequency sounds would be indicative of the effect of a hall on the tone quality of the sounds that the audience hears. For example, the effect of a hall on the perceived strength of the bass sounds should relate to low-frequency relative sound levels and the perceived strength of treble sounds to higher frequency relative sound levels. These effects might be like a simple tone control on a home stereo or perhaps we must consider the separate effects of early and late arriving sounds. Many texts suggest that for optimum conditions the reverberation time should increase at lower frequencies to provide an adequate sense of bass. Beranek proposed the bass ratio as a measure of bass balance [14]. This is the ratio of low-frequency to mid-frequency reverberation times. Barron's survey of British halls asked subjects to rate the bass and treble balance during concerts. Ratings of the bass balance were only significantly related to ratios of low- to mid-frequency early decay times and centre times [15]. He found no correlation with reverberation times or with Beranek's bass ratio of reverberation times. Barron found no objective correlates of treble balance.

Two recent subjective studies have shed a little more light on this topic. In the first, subjects rated the relative strength of bass and treble sounds in binaurally simulated sound fields. Binaural impulse responses obtained in concert halls were convolved with an anechoic music sample and were played back to subjects with a system that used near field loudspeakers [16]. Subjects listened to 10 different sound fields which had been selected to include a wide range of conditions typical of real halls. Subjective ratings of both perceived bass and treble were obtained.

The mean ratings of the perceived strength of treble sounds were related to what can be called a treble level balance. This was the difference between the 4 kHz and mid-frequency late arriving relative sound levels (*GL*(4k)–*GL*(1-2k)). That is, this measure indicates how much the late arriving 4 kHz sound levels exceed the mid-frequency values. The mean subjective ratings of perceived treble from this study are plotted versus this treble level balance in Figure 6. This high- to mid-frequency level balance is perhaps an obvious measure of the perceived strength of treble sounds but the choice of late arriving rather than early or total relative levels needs some further explanation. The reason that the late levels were better predictors of subjective responses is simply that it is the high-frequency late levels that vary most among halls. This is because the physical differences among halls that influence treble sounds have a greater effect on later arriving sound energy. Typically treble sound levels will vary due to differences in the amount of porous absorbing material in halls. These differences will most affect the later arriving sound energy, and hence subjective ratings are more strongly correlated with these differences in later arriving sound levels.

The perception of bass sounds is influenced by different factors. The same study using binaurally simulated sound fields found perceived bass strength to be best related to the relative level of early arriving low-frequency sounds. Figure 7 plots the mean subjective ratings of bass strength versus the relative levels of the early arriving

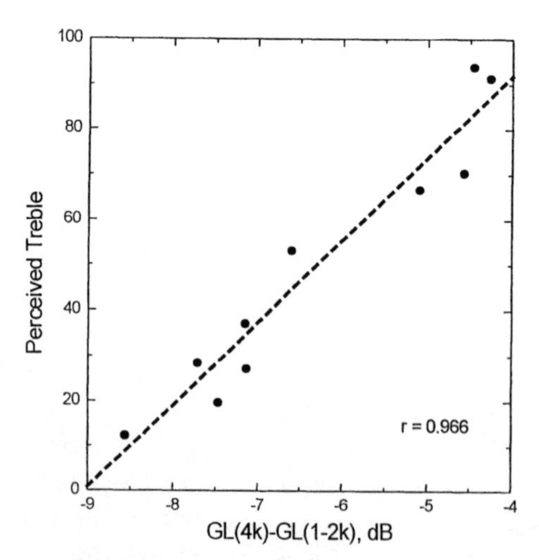

Figure 6. Mean perceived treble strength versus treble level balance of the late arriving sound [16].

sound summed over the octave bands from 125 to 500 Hz. Again the measure that best correlates with subjective ratings is the one that varies most among halls. Low-frequency sound levels vary most due to variations in the low-frequency seat-dip attenuation. The direct sound and some early reflections passing at near grazing incidence over the audience area are strongly attenuated at lower frequencies [11–13]. This phenomenon can produce a broad attenuation dip in the frequency response of the early arriving sound, and the centre frequency of this dip can vary from one location to another (see also section 2.4 below).

The effect of the seat-dip attenuation on bass sounds has been known for some time and one study suggested compensating for the reduced early arriving bass sound by adding more later arriving bass sound [17]. Some might conclude that this would require increased reverberation times at lower frequencies. While we do not know how to completely eliminate the seat dip attenuation, the magnitude of the effect has been related to ceiling height and adding non-grazing incidence low-frequency reflections such as those from lower ceilings might reduce the magnitude of this attenuation of bass sound.

The second subjective study systematically investigated the importance of various factors on the perceived strength of bass sounds [18]. Subjects listened to music in sound fields simulated using a computer controlled electro-acoustic system in an anechoic room. The low-frequency relative levels of the early and late arriving sound were varied independently as well as the low-frequency reverberation time and the direction of arrival of early arriving reflections. The range of variations in each acoustical measure was set to be as large as found in measurements of actual concert halls. For example, early arriving 125 Hz relative levels were varied over a 12 dB range, and 125 Hz reverberation times were varied from 1.4 to 3.2 s.

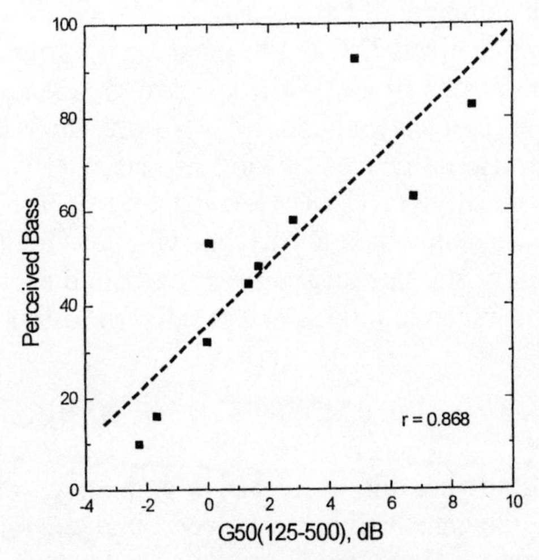

Figure 7. Mean perceived bass strength versus early bass level [16].

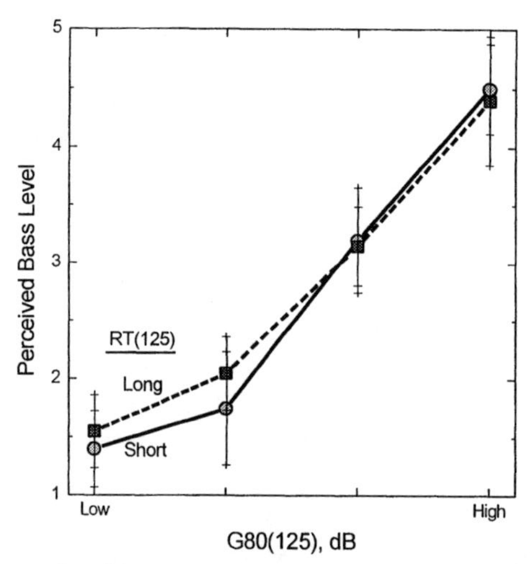

Figure 8. Mean perceived bass level as a function of early low-frequency sound level and low-frequency reverberation time [18].

Figure 8 shows the results of an experiment in which subjects rated the perceived bass strength of eight sound fields, consisting of the combinations of four levels of low-frequency early arriving sound and two levels of low-frequency reverberation time. The mean perceived bass strength increased with increasing early low-frequency sound level (*G80*(125)). This agrees with the ratings of binaurally simulated sound fields and again suggests that the seat-dip attenuation will influence perceptions of bass in real halls. However, these results also show that the low-frequency reverberation time had no significant effect on judgments of the strength of bass sounds. Increasing 125 Hz reverberation time from 1.4 to 3.2 s did not lead to any perceived increase in the strength of bass sounds.

In a second experiment, subjects listened to music in 12 sound fields consisting of the combinations of four levels of early arriving low-frequency sound and three levels of late arriving low-frequency sound. Both early and late arriving low-frequency sound significantly influenced mean responses. The perceived bass strength scores were more sensitive to 125 Hz levels than to 250 Hz levels and were more sensitive to later arriving sound than to early arriving low-frequency sound. The following combination of early and late arriving relative sound energies was found to be well correlated with the subjective ratings and is referred to as a weighted relative level, G_w,

$$G_w = 10\ \log\{E80(125)+3EL(125)+0.5[E80(250)+3EL(250)]\},\ \text{dB} \qquad (5)$$

where *E80*(125) and *E80*(250) are the early arriving relative sound energies and *EL*(125) and *EL*(250) are the late arriving relative sound energies. (e.g. $E80 = 10^{(G80/10)}$).

Figure 9 shows that there is a good relationship between the subjective ratings and these weighted relative levels. Thus perceptions of bass are influenced by both early and late arriving low-frequency sound. Since we are apparently more sensitive to differences in late arriving low-frequency sound, we can expect to be able to compensate for the attenuation of the early arriving sound by adding later arriving low-frequency sound as previously suggested [17].

When the direction of arrival of early reflections was varied, there were no systematic variations in the perceived strength of the bass sounds. This suggests that it is quite effective to compensate for the attenuation of grazing incidence low-frequency sound by adding low-frequency reflections from other directions such as from overhead. The studies using binaural simulations of sound fields in real auditoria indicate that the grazing incidence seat-dip attenuation does influence our perceptions of bass in real halls. However, the further studies of our sensitivity to bass sounds indicate that we can compensate for the seat dip attenuation by adding early reflections from other directions and by adding late arriving low-frequency sound.

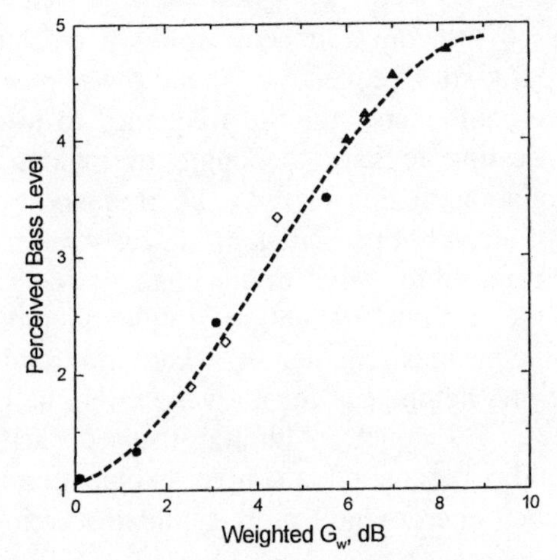

Figure 9. Perceived bass level versus weighted relative level G_W for varied early and late bass relative levels [18].

2.4 Designing for increased bass levels

The results in Figure 4 indicate that 125 Hz *G80* values are, on average, 3 dB less than would be expected from the reverberation time and room volume. Thus, on average, the seat-dip attenuation seems to reduce early arriving low-frequency sound by about 3 dB. In some halls and at some locations the effect can be considerably larger. We now know that this reduction in low-frequency sound levels will translate directly to the perception of reduced strength of bass sounds. This section considers the question of how we can most effectively increase low-frequency sound levels in halls that are lacking in bass strength.

The conventional approach would be to attempt to increase low-frequency sound levels by decreasing low-frequency sound absorption and hence increasing low-frequency reverberation time. A simple example calculation suggests that this approach can at best only lead to small improvements. If we use the Sabine reverberation time equation, we would find that an occupied hall with a volume of 20,000 m^3 and a 2 s reverberation time would have a total sound absorption of 1610 m^2. If there were 2,000 people in the hall, almost all of this absorption would be due to the audience. It is very unlikely that it would be possible to reduce the mount of low-frequency sound absorption by one-half in order to increase sound levels by the 3 dB that is, on average, lost to the seat dip attenuation.

A more effective approach would be to modify the seat-dip attenuation. Previous studies have shown that the amount of early arriving low-frequency sound is influenced by the ceiling height of the hall and that the frequency at which the maximum attenuation occurs can vary with the angle of incidence of the sound. One could effectively vary the angle of incidence by using more steeply raked seating. Unfortunately, varying the angle of incidence only varies the frequency at which the maximum attenuation occurs. The attenuation dip shifts in frequency but is not reduced in magnitude. Figure 10 from [13] shows that at near grazing incidence (grazing angle of incidence ≈ 0°) the dip frequency would be close to 200 Hz. As the angle of incidence was increased, the frequency of this dip was in some cases reduced to below 100 Hz. Thus, we could 'tune' the dip frequency to lower frequencies by using more steeply raked seating to vary the angle of incidence but we cannot eliminate the seat-dip attenuation in this way. (The frequency of the dip is also influenced by the direction of arrival of particular strong early reflections [19].)

Data from the 35 halls referred to earlier in this chapter were used to re-examine the effect of the ceiling on the seat dip attenuation. Figure 11 plots the hall average 125 Hz *G80* values versus the mean ceiling height. There is a trend for these levels to decrease with increasing ceiling height, but for a given ceiling height there is a wide range of these low-frequency *G80* values. Although these 125 Hz *G80* values may vary because of varying seat dip attenuation, a simpler explanation might be that halls with higher ceilings tend to be larger, to have more sound absorption and hence lower sound levels.

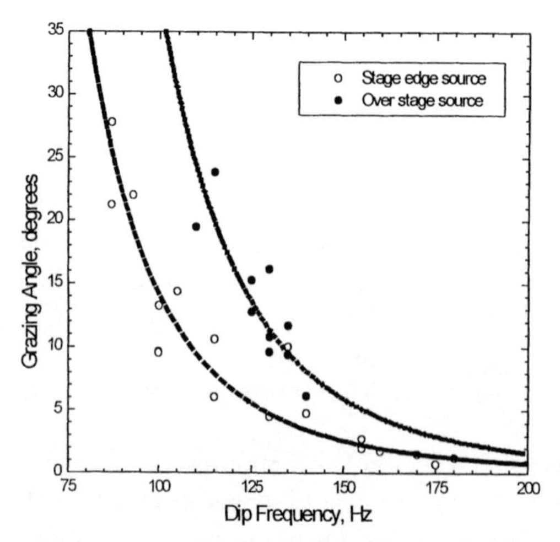

Figure 10. Variation of the frequency of maximum attenuation due to the seat dip effect as a function of the grazing angle of incidence.

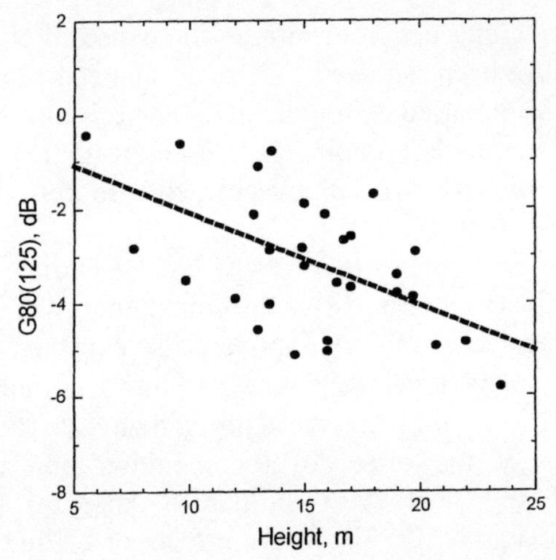

Figure 11. Mean measured 125 Hz *G80* values versus hall ceiling height.

Further calculations were made to estimate and remove the effect of reverberation time and sound absorption on these results. Barron's revised theory was used to estimate the expected 125 Hz *G80* values. These estimates give the estimated *G80* values based on reverberation times and room volumes but would ignore any influence of the seat dip effect. Normalized 125 Hz *G80* values were then created by subtracting the calculated estimates from the measured values. The resulting normalized 125 Hz *G80* values shown in Figure 12 more directly show the effects of the ceiling on average measured *G80* values. To further explain the influence of the ceiling, the ceilings were grouped into three categories. Those that would seem to provide especially diffuse reflections and those with over-stage reflectors providing stronger ceiling reflections were separated from the others. The normalized 125 Hz *G80* values in Figure 12 again show a trend for low-frequency sound levels to

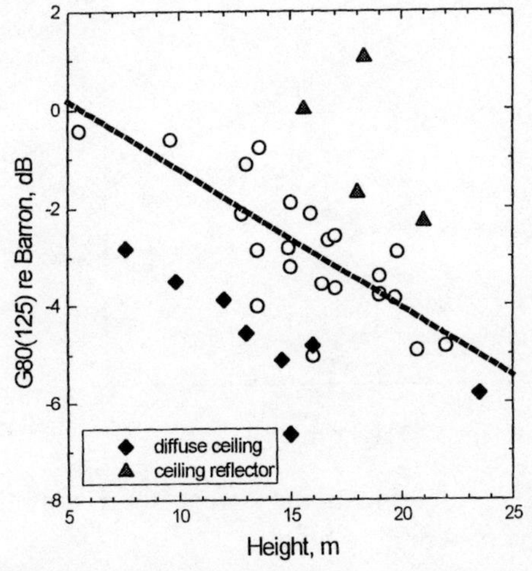

Figure 12. Normalized 125Hz *G80* values (with the effects of reverberation time and room volume removed) versus ceiling height.

decrease with increasing ceiling height even after the expected effects of hall volume and reverberation time have been removed. These normalized 125 Hz *G80* values can vary by more than 3 dB with varied ceiling height. There is also an obvious effect of the ceiling type. For intermediate ceiling heights (e.g. 15 m) these 125 Hz *G80* values can vary by as much as 4 dB between a diffuse ceiling and one with an effective ceiling reflector.

The influence of ceiling design and height on the total low-frequency sound levels was also examined. Because most of the sound energy is in the initial part of the impulse response, the effects of the ceiling were also expected to be significant for the total low-frequency sound levels. Figure 13 shows normalized 125 Hz total relative sound levels. That is, they are the hall average 125 Hz *G* values with the Barron theory estimates of the effect of reverberation time and room volume subtracted. The overall trends are very similar to those for early arriving low-frequency sound level in Figure 12. Both the effects of ceiling height and ceiling design are again present.

The average effect of the seat dip attenuation reduces early arriving sound levels by about 3 dB. Reducing low-frequency sound absorption to increase low-frequency sound levels could at best only partially compensate for this effect. However, the height and design of a ceiling can lead to variations in early arriving low-frequency sound levels that could be this large or even larger. Thus, hall design must consider the effect of the hall ceiling on low-frequency sound levels and the perceived strength of bass sounds in the hall.

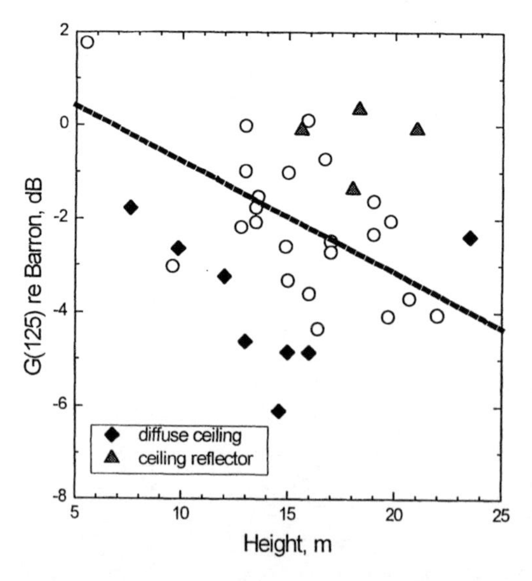

Figure 13. Normalized 125 *G* values (with the effects of reverberation and room volume removed) versus ceiling height.

3 Spatial impression in halls

3.1 Some history

Spatial impression refers to how the acoustical impression in a room or space is different from that outdoors or some other situation without many delayed reflections arriving from a variety of directions. Early references related spatial effects to reverberation [20] because reverberation is perhaps the most obvious effect of large spaces. In the 1960s it was suggested that spatial impression was related to early arriving lateral reflections. Work by Barron and Marshal confirmed this and related spatial impression to the fraction of the early sound that arrived from lateral directions [5,6]. This new understanding of the importance of early lateral reflections significantly advanced the science of concert hall acoustics and has had a great impact on hall design [21,22]. More recently, it has gradually emerged that spatial impression has at least two separate dimensions [23–25]. The apparent source width, ASW, is related to the strength of early lateral reflections and listener envelopment, LEV, is related to the strength of late arriving lateral sound energy. Thus, we can now look back and say that Barron and Marshall identified the ASW component of spatial impression. Because LEV is related to the late arriving lateral energy there are some connections back to reverberation time.

3.2 Apparent source width (ASW) and listener envelopment (LEV)

Barron's pioneering studies of spatial impression related what we now call ASW to the lateral energy fraction [5,6]. The lateral energy fraction, *LF*, is the fraction of the early energy that arrives from the side. *LF* is defined as follows,

$$LF = \int_{0.005}^{0.08} p_L^2(t)dt \Big/ \int_{0}^{0.08} p^2(t)dt \tag{6}$$

where $p(t)$ is the instantaneous pressure in the halls impulse response and $p_L(t)$ is the instantaneous pressure response of the lateral energy which was defined in terms of a cosine directivity pattern. Barron suggested that *LF* values should be measured over the octave bands from 125 to 1000 Hz.

Figure 14 shows recent data that confirm the results of the earlier studies of early lateral reflections [26]. Subjective ratings of the apparent source width from a simple paired comparison test increased with increasing *LF* values. This graph also indicates the important effect of sound level on ASW. That is, both the level of the sound and the fraction of early energy that arrives from lateral directions are critical to creating the impression of source broadening. It has been suggested [27] that the relative level of the early lateral sound might be an effective single measure of both of these effects but the idea has not been thoroughly evaluated.

The apparent source width has also been related to inter-aural cross-correlation measures [9,26,28]. Although these measures would appear to be quite different from lateral energy fractions, inter-aural cross-correlations of the early part of binaural impulse responses, *IACC(E)*, have been related to *LF* values. Since the inter-aural

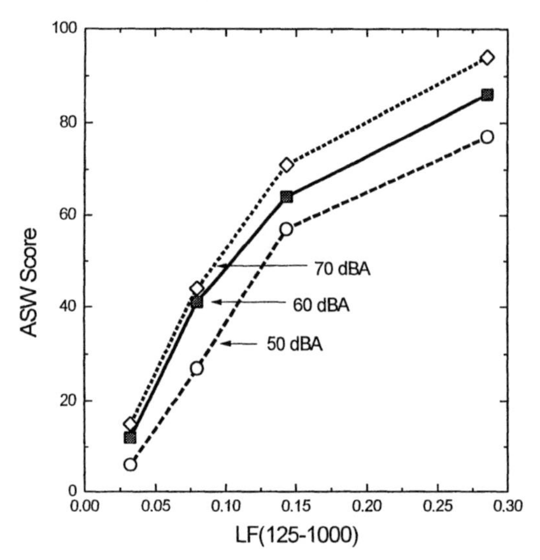

Figure 14. Mean subjective rating of ASW versus *LF*(125-1000) for three sound levels.

cross-correlations are related to the similarity of the signals at the two ears, *LF* values relate directly to 1–*IACC(E)* values. Barron suggested an approximate theoretical relationship [29], and recent experimental studies [30] have demonstrated the two types of measures to be related.

From experimental studies in auditoria, hall average *LF* values were significantly correlated with (1–*IACC(E)*) for the octave bands from 125 to 1000 Hz[30]. (*IACC(E)* is the inter-aural cross-correlation of the early arriving sound.) These results are repeated here in Figure 15. In the 125 Hz octave the 1–*IACC(E)* values do not vary as much as in the mid-frequency octave bands, but the 1–*IACC(E)* values are strongly correlated with *LF* values at 125 Hz. However, in the highest two octave bands the two measures are less well correlated. At these frequencies, details of the dummy head used to make the measurements, and the direction of arrival of particular early reflections are expected to influence *IACC(E)* values but not *LF* values and hence reduce the correlation between the two measures.

When measurements at individual positions are considered, there is greater scatter in the plots comparing the two types of measures but the two measures are again best correlated at mid-frequencies. When considered in detail, the two types of measures are apparently influenced differently by various factors at particular locations. However, for the four octave bands most important to spatial impression (125–1000 Hz), the hall average values of *LF* and *IACC(E)* values are strongly related and hence provide essentially the same information about the average properties of a hall.

The early subjective studies of spatial impression and those shown here in Figure 14 were based on very simple simulated sound fields. They include only a direct sound and two early reflections. These were quite simplistic, and recent studies [25] have shown that using more realistic sound fields, including the direct sound and a

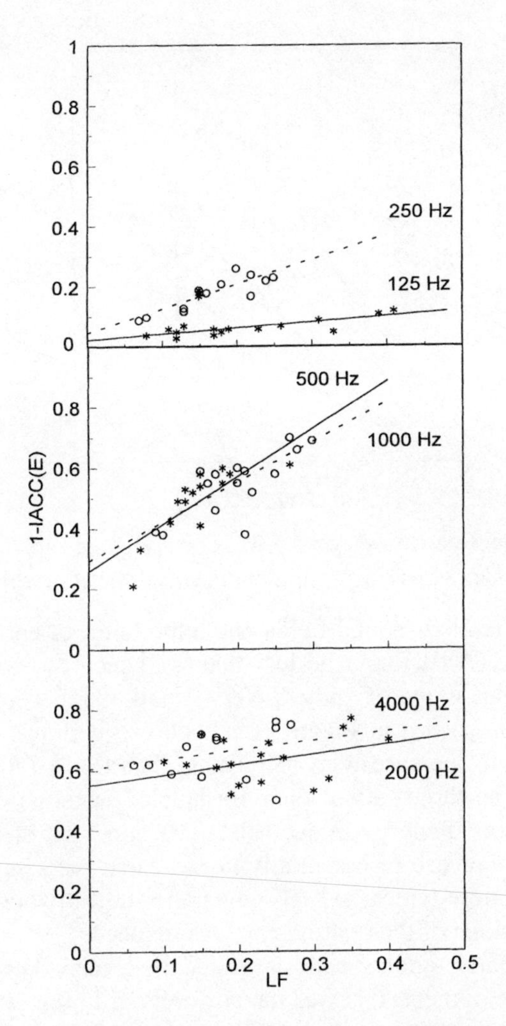

Figure 15. Hall average values of 1–*IACC(E)* versus *LF*.

sequence of early reflections followed by a reverberant tail, leads to different results. The effect of adding later arriving sound reduces the perceived source broadening caused by the early lateral reflections.

In a series of experiments, subjects listened to music reproduced via simulated sound fields. The sound fields were simulated using a computer controlled electro-acoustic system in an anechoic room. Subjects compared pairs of sound fields and rated the differences on a five-point scale. In the first experiment, subjects rated the difference in ASW between the sound field pairs. When only early reflections were included, quite large variations in ASW were reported. When later arriving sound energy was added, the subjective ratings of differences in ASW were reduced even though the early reflections remained unchanged. Thus, adding late arriving sound seemed to reduce subjects' sensitivity to the effects of early lateral reflections. This

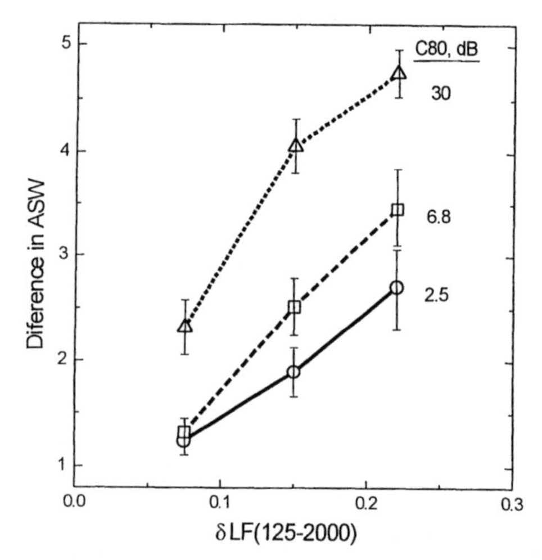

Figure 16. Mean subjective ratings of ASW versus changes in *LF* for three different amounts of late arriving sound energy measured in terms of *C80*.

suggests that, in more realistic sound fields, the importance of early lateral reflections and source broadening (ASW) may be less than indicated by experiments based on simple sound fields made up of only a very small number of reflections. The experimental results are shown in Figure 16. Each curve shows mean judgments of differences in ASW to increase with increasing changes in *LF* between the sound fields of each pair. The three sets of data are labeled in terms of *C80*, the ratio of early-to-late arriving sound energy in decibels. The upper set of data, with a *C80* of 30 dB, correspond to the case of essentially no late arriving sound energy. As *C80* was reduced to values more typical of real rooms, the subjective ratings of ASW also were reduced and the slope of the relationship was reduced.

Although some early studies had suggested that early lateral reflections also created a sense of envelopment, this was never observed in this series of experiments and further investigations focussed on the conditions necessary for a sense of envelopment. A sense of listener envelopment LEV was only detected when later arriving reflections arriving from the side were present. Several experiments were designed to determine the aspects of the late arriving sound energy that contribute to the sense of LEV.

To demonstrate that LEV occurred only when late arriving energy was present, a digital reverberator that provided a gated burst of energy was used. The burst duration was 80 ms and the start of the burst relative to the arrival of the direct sound was varied. As illustrated in Figure 17, mean judgments of LEV increased as the start of the gated burst was delayed. When the burst started at 0 ms after the direct sound, perceived LEV was minimal because in this case all of the energy arrived within the first 80 ms. Figure 17 also shows that when the level of the burst was increased, LEV was also increased. Thus, a sense of envelopment increases with increasing late arriving sound and with increasing delay of the time of arrival of this sound.

When sound fields are more reverberant, more sound energy tends to arrive later in time after the direct sound. Therefore, one would expect LEV to increase with

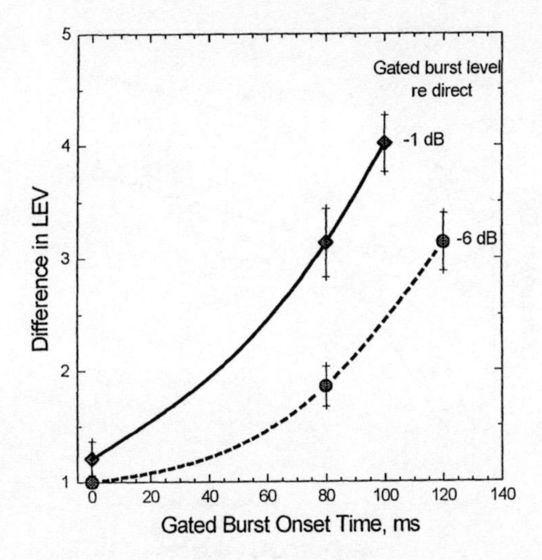

Figure 17. Mean subjective ratings of listener envelopment versus the start time of a gated burst of sound and for two different levels of late sound.

increasing reverberation time. The next experiment included more realistic sound fields with a sequence of early reflections followed by a complete exponential decay. In the same experiment the level of the late arriving sound was also varied and this was described in terms of the early-to-late sound ratio, *C80*. The results in Figure 18 show that mean LEV scores increase with decreasing *C80*, corresponding to increasing levels of late arriving sound. As expected, LEV also increased with increasing reverberation time. Thus, increased late arriving sound energy that is delayed more in time leads to an increased sense of envelopment.

Listener envelopment was also found to depend on the direction of arrival of the late lateral sound energy. Subjects listened to music via simulated sound fields with

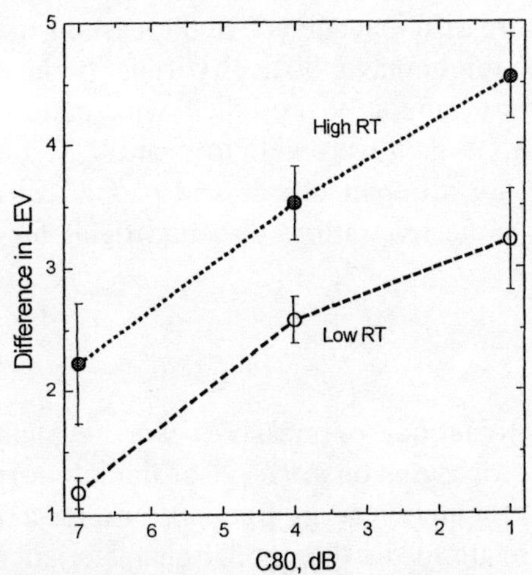

Figure 18. Mean subjective rating of LEV versus early/late sound ratio (*C80*) for two different reverberation times.

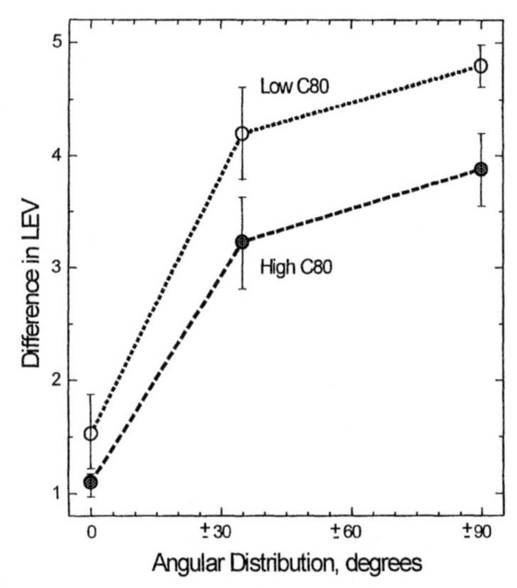

Figure 19. Mean LEV versus angular distribution of the late arriving sound for two levels of late energy.

three different types of angle of arrival and two different late sound levels. The late sound energy was spread over angles of: 0 degrees (from only straight ahead), ±35 degrees, and ±90 degrees. The mean subjective ratings of *LEV* in Figure 19 indicate that there is essentially no envelopment when the late sound arrives from straight ahead (0 degrees). However, when the late arriving sound arrives from an angle of ±35 degrees, quite strong LEV is noted. Increasing the angles of arrival to ±90 degrees further increased the sense of LEV. Thus, a sense of envelopment occurs when there is sufficient late arriving sound energy arriving from the side of the listener. Increasing the level of the late energy, delaying the time of arrival and increasing the angular spread of the late arriving energy all lead to an increased sense of LEV.

While ASW and LEV are different subjective dimensions of spatial impression, it was thought that listeners might have difficulty reliably identifying each effect. However when subjects were randomly presented with pairs of sound fields with either changes in LEV or ASW they were able to identify 81 to 94% of the changes correctly. LEV and ASW are different effects and people can readily detect them. They are both important to subjective ratings of sound fields for audiences in concert halls.

3.3 Late lateral sound levels

Various measures and combinations of measures were evaluated as predictors of listener envelopment [31]. Measures of the level of the late arriving sound relate to the level component of LEV. Lateral energy fractions or inter-aural cross-correlations of the late arriving sound relate to the directional characteristics of the late arriving sound energy. Decay times are indicative of the temporal characteristics of the sound field with longer decay times indicating energy delayed more in time. Combinations of levels and lateral fractions or levels and inter-aural cross-correlation were found to correlate well with LEV ratings. However, the best correlation with subjective ratings of LEV was an elegantly simple measure, the late lateral relative sound level, *GLL*. This is a relative sound level of only the late arriving energy as measured using a figure-of-eight pattern microphone to pick up only lateral energy. It was best correlated with LEV scores when the *GLL* values were summed over the octaves from 125 to 1000 Hz. Figure 20 shows the highly significant relationship between mean LEV scores and *GLL* values summed over the octaves from 125 to 1000 Hz.

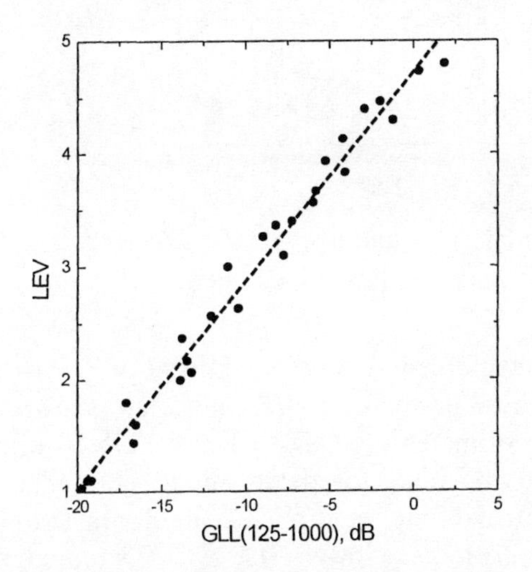

Figure 20. Mean subjective ratings of LEV versus the late lateral sound level *GLL* [31].

3.4 The relation between late lateral sound levels and hall shape

Knowing that late lateral sound levels *GLL* are indicative of the degree of perceived listener envelopment, we can now use this measure to evaluate the degree of envelopment in concert halls. By comparing *GLL* measurements with the geometry and other physical aspects of halls, we can hope to learn how to design halls that have appropriately enveloping sound fields. *GLL* values were calculated for 16 different halls, and the hall average values are plotted in Figure 21. The hall average *GLL* values vary over a 12 dB range and so this measure does vary significantly among halls. By ordering the halls in terms of mean *GLL* value in this figure, it is seen that there is a trend for rectangular-plan halls to have higher *GLL* values and fan-shaped halls to have lower *GLL* values. A number of factors may combine to produce higher *GLL* values. For example, halls with longer reverberation times will tend to have higher sound levels and hence higher *GLL* values. To better identify the influence of hall shape, normalized *GLL* values were calculated with the effects of reverberation time and room volume on late arriving sound levels removed. This was done by subtracting calculated late sound levels (using Barron's theory, equation (4)) from the mean measured *GLL* values and normalizing the results to a mean value of 0 dB. This

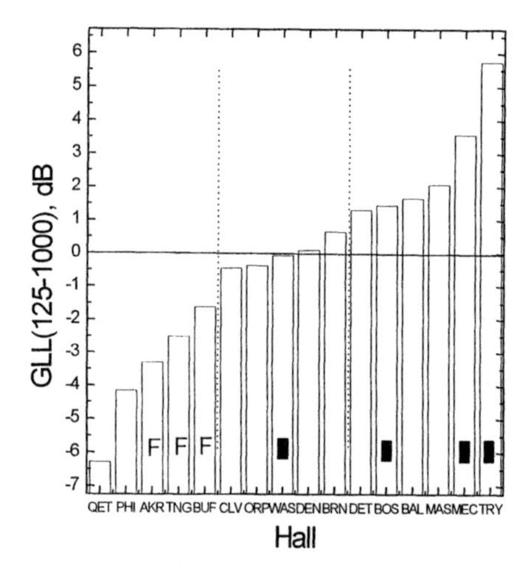

Figure 21. Hall average late lateral sound levels, *GLL*. Rectangles indicate rectangular plan halls and 'F' indicates fan shaped halls.

was thought to be a reasonable estimate of the effects of reverberation and room volume on *GLL* values. These normalized *GLL* values are shown in Figure 22 and again are presented in order of increasing *GLL*. When the effects of reverberation and room volume are removed the range of the resulting normalized *GLL* values is reduced to about 4 dB. However, the trend for rectangular halls to have higher *GLL* values and for fan shaped halls to have lower *GLL* values is more obvious. It is clear that rectangular halls naturally tend to have more enveloping sound fields, and fan shaped halls will tend to be lacking in LEV. However, hall shape may further influence envelopment in that rectangular halls appear to be likely to have higher late lateral sound energy due to higher reverberation times.

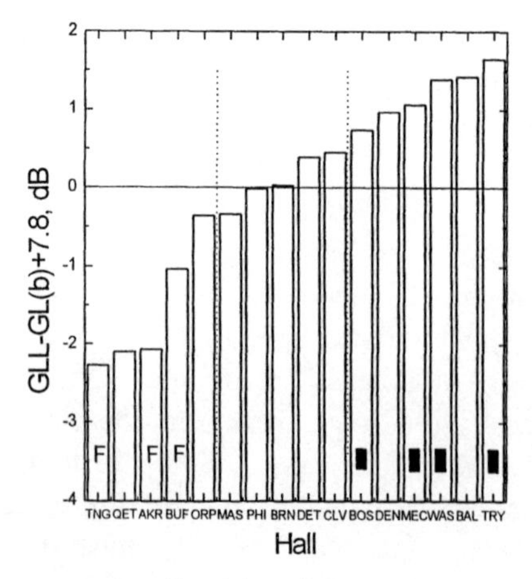

Figure 22. Hall average normalized late lateral sound levels, *GLL*.-*GL*(*b*). Rectangles indicate rectangular plan halls and 'F' indicates fan shaped halls, (where *GL*(*b*) is the late level from eqn (4)).

3.5 Spatial impression under-balconies

Our new understanding of spatial impression helps to more completely explain conditions that occur at under-balcony locations. It is often quite obvious to listeners that acoustical conditions under large balconies are quite different than in other parts of the same hall. It has often been suggested that spatial impression is different at under-balcony seats. Until recently, differences in spatial impression would be expected to indicate differences in early reflections. However, as the results in Figures 23 and 24 indicate, early sound levels do not appear to be systematically influenced by the presence of the balcony but late arriving sound levels are reduced at under-balcony seats.

Figure 23 plots early and late relative sound levels versus distance for 1000 Hz results in Boston Symphony Hall. This hall has only a very modest balcony overhang but late arriving sound levels are apparently reduced at under-balcony seats. Early arriving relative sound levels show no particular effect of the balcony. Both early and late measured sound levels vary with distance approximately as predicted by Barron's theory shown by the dashed line on these plots.

The effects of the much larger balcony overhang in the Orpheum Theatre are illustrated in Figure 24. Again, there is no systematic effect of the balcony on early arriving sound. There is a quite large scatter in the plot of early levels versus distance because when measured the ceiling caused strongly focussed reflections at some locations. (These problems have since been rectified [32].) The late arriving sound levels are quite obviously reduced at under-balcony seats in this hall.

In both cases, with either a small or a large balcony overhang, the effect of the balcony is to reduce late arriving sound energy compared to seats at similar distances but not under the balcony. This is in complete agreement with a recent study by Barron[33].

If we combine this with our new understanding of spatial impression the picture becomes complete. There is indeed reduced spatial impression at under-balcony seats. However, this is not reduced ASW but reduced LEV because it is the late energy and not the early energy that is modified by the balcony. This can be further demonstrated by examining late lateral relative levels plotted versus distance for the Orpheum Theatre in Figure 25. As expected, *GLL* is clearly reduced at under-balcony seats compared to other locations at a similar distance from the source. We now know that *GLL* is a strong correlate of LEV and hence can confirm that sound fields at under-balcony seats are lacking in envelopment.

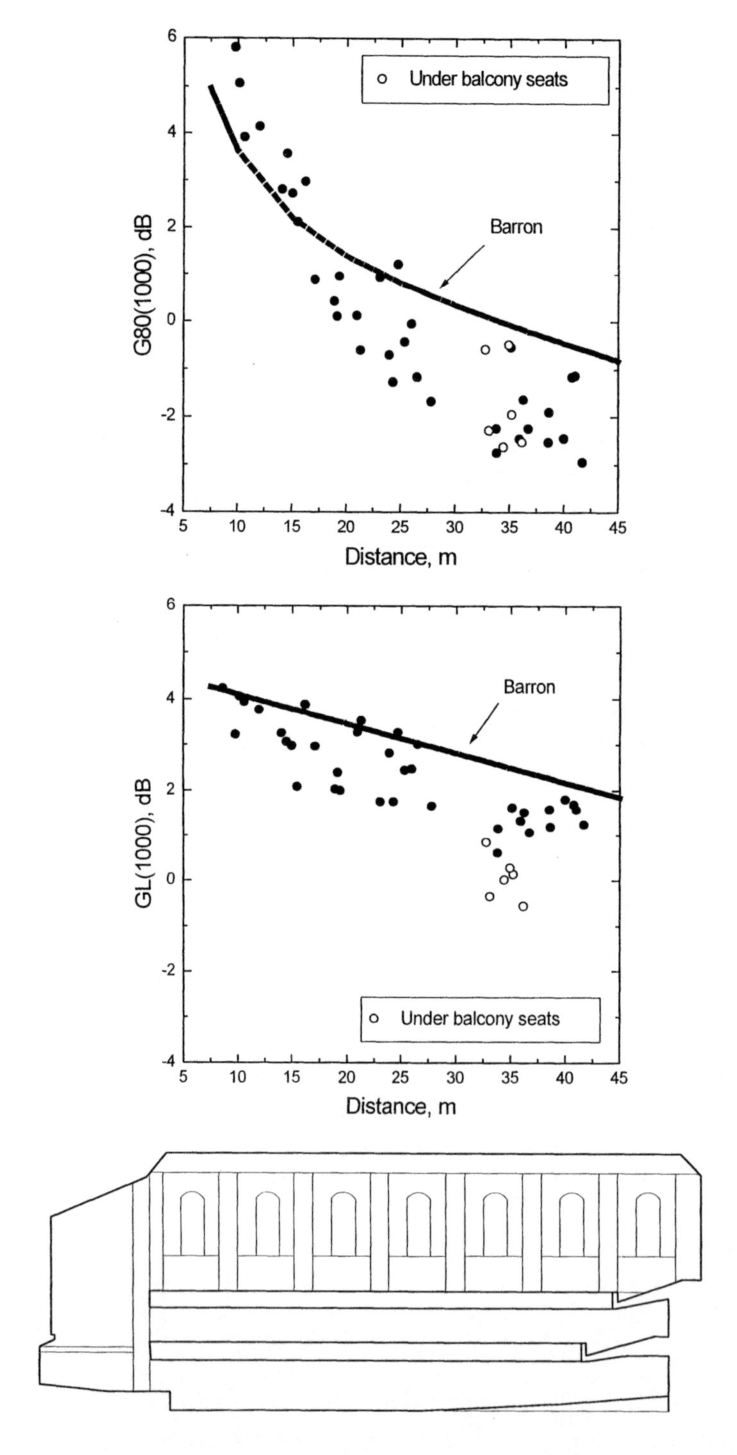

Figure 23. Boston Symphony Hall, (top) relative early sound level *G80* versus distance, (middle) relative late sound level *GL* versus distance, and (lower) section of the hall showing the small balcony overhang.

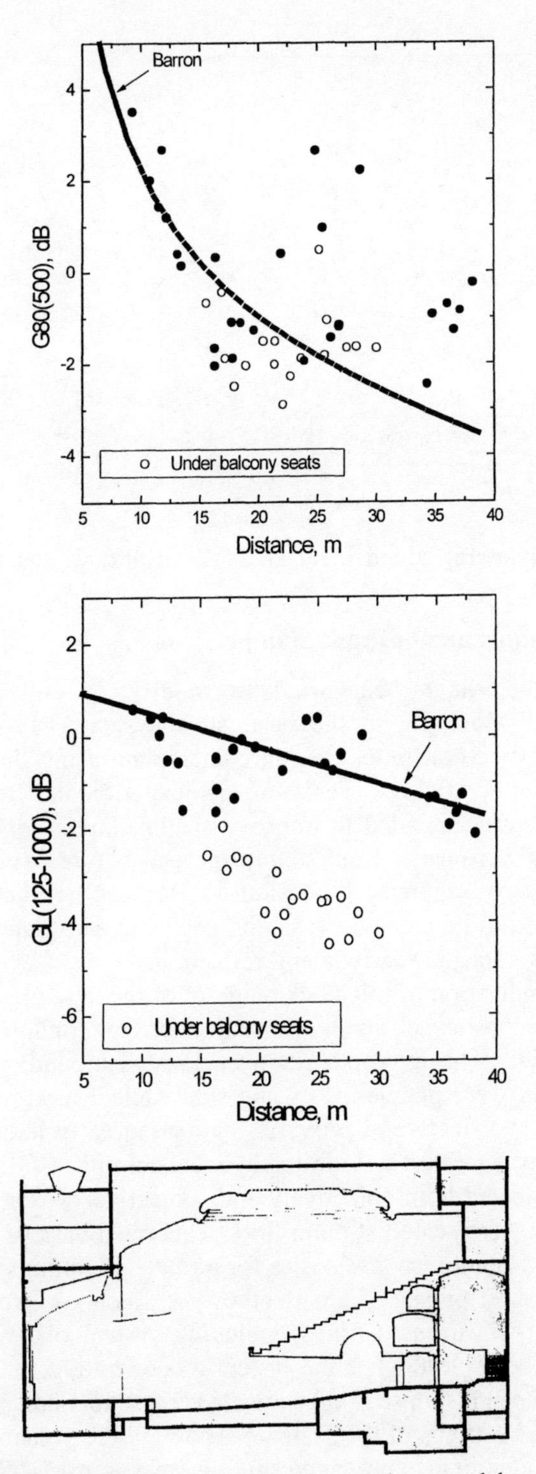

Figure 24. Orpheum Theatre, (top) relative early sound level *G80* versus distance, (middle) relative late sound level *GL* versus distance, and (lower) section of the hall showing the large balcony overhang.

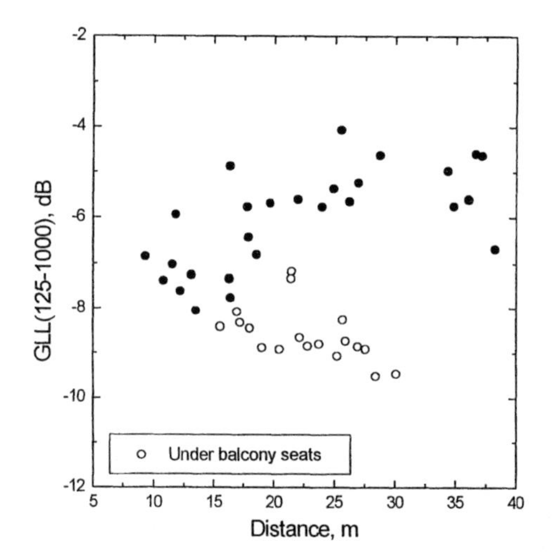

Figure 25. Relative late lateral sound level $GLL(125\text{–}1000)$ versus distance in the Orpheum Theatre.

3.6 Electro-acoustic enhancement of spatial impression

Electro-acoustic techniques have been used to modify or enhance acoustical conditions in auditoria. Although, in the past, such systems have usually been designed to enhance reverberation time, we can contemplate using them to enhance other aspects of the sound field. Tests of an experimental electro-acoustic enhancement system that was intended to improve spatial impression are described below. The experiments were carried out some time ago before it was known that spatial impression had two separate dimensions. Because at that time spatial impression was understood to be created by having adequate early lateral reflections, the system was designed to enhance early lateral reflections.

Figure 26 gives an indication of the block diagram of the system. It consisted of 20 independent electro-acoustics channels consisting of: microphone, time delay, amplifier and loudspeaker. The microphones were located around the proscenium arch and the loudspeakers were positioned on the side wall of a large multipurpose hall. The system was very effective at providing increased early lateral reflections. The useful effects of the system were not limited by feedback.

The system was evaluated both objectively and subjectively. For the subjective evaluations, subjects who were seated at main floor seats rated various aspects of the acoustical experience on 5-point response scales for a range of settings of the system. An orchestra repeated a short passage from Beethoven's Leonore Overture for each setting of the system. The various settings, including system off, were randomly presented and subjects were not told when the system was operating.

Figure 27 plots the mean response of 50 subjects versus the mean measured early decay times for the various settings of the system. Although the system was intended to increase early lateral sound, it also produced small increases in *EDT* values of up to about 0.3 s. The results in Figure 27 show that subjective ratings of reverberance were significantly related to the mean *EDT* values and confirmed that people could reliably detect these small improvements.

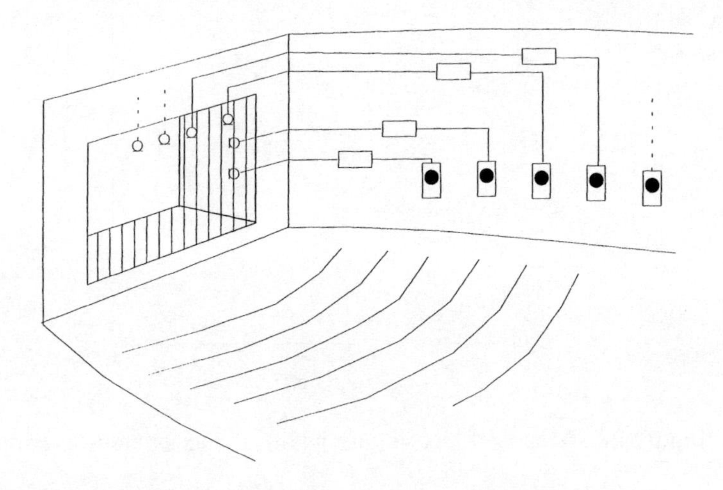

Figure 26. Block diagram of 20 channel electro-acoustic enhancement system.

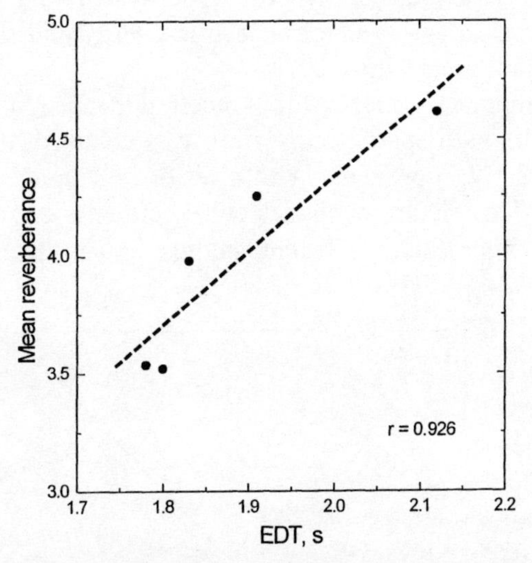

Figure 27. Mean subjective ratings of reverberance versus average *EDT.*

Although the system increased overall sound levels by up to only 2 dB, Figure 28 shows a strong correlation between subjective judgments of loudness and the average measured relative sound levels. Again subjects could, on average, reliably detect these small increases in sound level produced by the system.

As intended, the system was quite effective at increasing early lateral sound. The

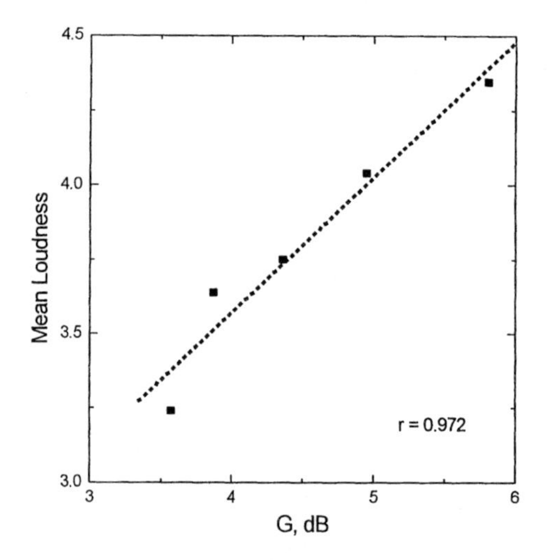

Figure 28. Mean subjective ratings of loudness versus average *G*.

results in Figure 29 show that the system was capable of doubling the measured lateral energy fraction. The figure also shows a strong correlation between judgments of spatial impression and measured *LF* values. The system did exactly what it was intended to do. It increased early lateral reflection energy and subjects clearly heard this as a change in spatial impression.

In spite of being apparently successful in meeting the design objectives, the result was disappointing. Although spatial impression was changed, the overall acoustical impression was not greatly improved. With a lot of hindsight, it is now possible to explain these results. This electro-acoustics enhancement system had only increased one aspect of spatial impression. Recent studies in simulated sound fields [25]

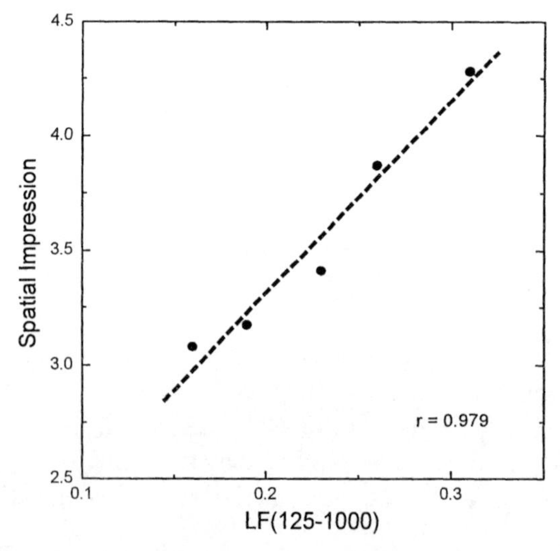

Figure 29. Mean subjective ratings of spatial impression versus average *LF*.

suggest that the other aspect of spatial impression, namely listener envelopment, is probably the more important of the two dimensions. We can now understand that the system would have been much more successful if it had been designed to enhance listener envelopment. Of course, this suggests that electro-acoustic enhancement systems do have great potential for enhancing more than just reverberation time. In particular, it should be easier to provide increased late lateral energy than enhanced early reflections.

4 Conclusions

The major conclusions of this chapter can be summarized by the following points:

- Barron's theory is demonstrated to be a useful and accurate tool for considering early, late and total sound levels in halls. It was shown that the accuracy of the Barron equations can be improved by basing the calculations on early decay time values rather than on reverberation times.

- Early low-frequency sound levels are, on average, 3 dB less than expected from reverberation time and room volumes due to the excess low-frequency attenuation of the seat dip effect.

- The low-frequency seat dip attenuation can be modified by changes to the height and reflecting qualities of the hall ceiling. Halls with lower ceilings tend to have stronger bass. Halls with diffusing ceilings tend to have weaker bass but halls with effective over-stage reflectors tend to have stronger bass sound.

- The perceived strength of treble sounds in halls is related to the treble level ratio of the late arriving sound.

- The perceived strength of bass sounds in halls is related to the early bass sound level.

- These physical measures of bass and treble strength are successful because they vary most among halls.

- The perceived strength of bass sounds in simulated sound fields is related to both early and late low-frequency sound levels but not to low-frequency reverberation times.

- We can compensate for the reduction of low-frequency energy caused by the seat dip effect by adding late arriving low-frequency sound or reflected sound from other directions that will not be attenuated by the seat dip effect.

- We now know that spatial impression has two subjective dimensions: apparent source width ASW and listener envelopment LEV. *ASW* is related to the level of early lateral reflections but *LEV* is correlated with the strength of later arriving lateral reflections.

- The late lateral sound level summed over the octaves from 125 to 1000 Hz, *GLL*(125–1000), is the best indicator of perceived listener envelopment.

- Sound fields under balconies are lacking in late arriving sound and hence there is a reduced sense of listener envelopment.

- There is a great potential to use electro-acoustic enhancement techniques to provide an increased sense of listener envelopment in halls.

Although these results represent further progress in the understanding and design of concert halls there is much more to be done. For example, studies of spatial impression have, to date, concentrated on reflections arriving in the horizontal plane of the listener's ears. There is very little information concerning the perception of sounds arriving from other directions. There is not even much information on the subjective effects of sound reflections arriving from behind the listener.

We have a number of physical measures that relate to various aspects of spatial impression as well as to the loudness and tone quality of sounds in halls. For example, we can now use late lateral sound levels to estimate the degree of listener envelopment or early bass levels to indicate the expected perceived level of bass sound. However, for these two examples and for most of the other measures discussed in this chapter, we have practically no information about optimum or even acceptable values of each of these measures. There is clearly much more to do.

References

[1] Sabine, W.C., *Collected Papers on Acoustics*, Harvard University Press (Reprinted by Acoustical Society of America, New York, 1993).

[2] Cremer, L., and Müller, H.A., *Principles and Applications of Room Acoustics*, Volume 1, Applied Science Publishers, London, 1978.

[3] Barron, M., *Auditorium Acoustics and Architectural Design*, E & FN Spon, London 1993.

[4] Beranek, L.L., *Concert Halls and Opera Houses: How They Sound,* Acoustical Society of America, New York, 1996.

[5] Barron, M., The subjective effects of first reflections in concert halls: the need for lateral reflections, *J. Sound Vibr.*, **15**, pp. 475-494, 1971.

[6] Barron, M., and Marshall, A.H., Spatial impression due to early lateral reflections in concert halls: the derivation of a physical measure, *J. Sound Vibr.*, **77**, pp. 211-232, 1981.

[7] Schroeder, M.R., Gottlob, D., and Siebrasse, K.F., Comparative study of European concert halls: correlation of subjective preference with geometric and acoustic parameters, *J. Acoust. Soc. Am.*, **56**, pp. 1195-1201, 1974.

[8] Lehmann, P., and Wilkens, H., The connection between subjective judgements of concert Halls, *Acustica*, **15**, pp. 256-268, 1980.

[9] Ando, Y., *Concert Hall Acoustics*, Springer-Verlag, Berlin, 1985.

[10] Barron, M., and Lee, L.-J., Energy relations in concert auditoria, I, *J. Acoust. Soc. Am.* **84**, pp. 618-628, 1988.

[11] Schultz, T.J., and Watters, B.G., Propagation of sound across audience seating, *J. Acoust. Soc. Am.*, **36**, pp. 885-896, 1964.

[12] Sessler, G.M., and West J.E., Sound transmission over theater seats, *J. Acoust. Soc. Am.* **36**, pp. 1725-1732, 1964.

[13] Bradley J.S., Some further investigations of the seat dip effect, *J. Acoust. Soc. Am.*, **90**, pp. 324-333, 1991.

[14] Beranek, L.L., *Music Acoustics and Architecture*, John Wiley & Sons, New York, 1962.

[15] Barron, M., Subjective study of British symphony concert halls, *Acustica*, **66**, pp. 1-14, 1988.

[16] Soulodre, G.A., and Bradley, J.S., Subjective evaluation of new room acoustic measures, *J. Acoust. Soc. Am.*, **98**, pp. 294-301, 1995.

[17] Schultz, T.J., Acoustics of the concert hall, *IEEE Spectrum*, pp. 56-67, 1965.

[18] Bradley, J.S., Soulodre, G.A., and Norcross, S., Factors influencing the perception of bass, *J. Acoust. Soc. Am.*, **101**, pp. 3135, 1997.

[19] Bradley, J.S., Some effects of orchestra shells, *J. Acoust. Soc. Am.*, **100**, pp. 889-898, 1996.

[20] Kuttruff, H, *Room Acoustics*, John Wiley and Sons, New York, p. 196, 1973.

[21] Marshall, H.A., The acoustical design of concert halls, *Inter-Noise 83*, pp. 21-30, 1983.

[22] Paoletti, D.A., Hyde, J.R., and Marshall, H.A., The acoustical design of the Orange County Performing Arts Center theater, *Proceedings of 11th International Congress on Acoustics*, Paris, pp. 157-160, 1983.

[23] Morimoto, M., and Maekawa, Z., Auditory spaciousness and envelopment, *Proceedings 13th International Congress on Acoustics*, pp. 215-218, Belgrade, 1989.

[24] Morimoto, M., and Iida, K., A new physical measure for psychological evaluation of a sound field; front/back energy ratio as a measure of envelopment, *J. Acoust. Soc. Am.*, **93**, p. 2282, 1993.

[25] Bradley, J.S., and Soulodre, G.A., The influence of late arriving energy on spatial Impression, *J. Acoust. Soc. Am.*, **97**, pp. 2263-2271, 1995.

[26] Bradley, J.S., Soulodre, G. A., and Popplewell, N., Pilot study of simulated spaciousness, *J. Acoust. Soc. Am.*, **93**, p. 283, 1993.

[27] Bradley J.S., Hall average characteristics of 10 halls, *Proceedings of 13th International Conference on Acoustics*, Belgrade Yugoslavia, **2**, pp. 201-202, 1989.

[28] Keet, W. de V., The influence of early lateral reflections on the spatial impression, *Proceedings of the 6th International Congress on Acoustics*, paper E-2-4, Tokyo, 1968.

[29] Barron, M., Objective measures of spatial impression in concert halls, *Proceedings of the 11th International Congress on Acoustics*, Paris, pp. 105-108, 1983.

[30] Bradley, J.S., Comparison of concert hall measurements of spatial impression, *J. Acoust. Soc. Am.*, **96**, pp. 3525-3535, 1994.

[31] Bradley, J.S., and Soulodre, G.A., Objective measures of listener envelopment, *J. Acoust. Soc. Am.*, **98**, pp. 2590-2597, 1995.

[32] O'Keefe, J., Soulodre, G., and Bradley, J., Acoustical renovation of the Orpheum Theatre, Vancouver Canada, *Proceedings of 135th ASA and 16th ICA*, Seattle, **4**, p. 2459, June 1998.

[33] Barron, M., Balcony overhangs in concert auditoria, *J. Acoust. Soc. Am.*, **98**, pp. 2580-2589, 1995.

Chapter 6

Acoustics in churches

J.J. Sendra, T. Zamarreño and J. Navarro
Escuela Técnica Superior de Arquitectura,
Universidad de Sevilla,
Avda. Reina Mercedes 2, 41012 Sevilla, Spain
Email: jsendra@cica.es

Abstract

Churches of outstanding historical and architectural quality in Western European cities are often renovated as theatres, concert halls and conference halls as a means of safeguarding this artistic heritage. However, the lack of an acoustic study in the renovation process can lead to serious functional problems. In this chapter, we would like to stress the special characteristics of this type of enclosure, which has played such an important role in the history of architecture. We will begin by evaluating the acoustic properties of the principal types of churches, from early Christian structures until the still recent Late Baroque churches. Then we will make an acoustic analysis of a model of a church that is widespread in southern Spain and that is characterized by good acoustic conditions: Mudejar–Gothic churches with timber roofs. This analysis also gives us the opportunity to present a methodology that our group has used to carry out various acoustic studies on churches while taking into account their spatial peculiarities. Finally, we will describe three church renovation projects we have participated in and the solutions used, which we think may be useful to other architects.

1 Introduction

This paper summarizes the research carried out in recent years on church acoustics. Some of this study's results and conclusions have been expounded in several scientific conferences or have been published, as we will later see.

A series of acoustic renovation projects carried out by our team in churches led us to write this article. With a few exceptions, we intervened at the request of government institutions that had substantiated the extremely poor acoustics of churches that had recently undergone a costly renovation process to adapt them to cultural use as theatres, concert halls, conference halls, etc. We must bear in mind that this type of church renovation (and subsequent new use) is part of a cultural policy that is firmly established in Western Europe to safeguard our rich architectural heritage.

These functional problems can be explained, in some cases, because the architects did not deem acoustics to be a fundamental aspect meriting a detailed study for the uses described. Also, the measures needed to improve a building's acoustics are not always simple and may require fundamental changes in the renovation project. In other cases, the resulting deficient acoustics came as a surprise to the architects because their plans proposed insubstantial solutions meant simply to recondition the church room, a place that had traditionally been "for listening to music" (Navarro and Sendra [1]) and had originally been designed to preach the "Word of God" (Sendra and Navarro [2, 3]).

In reality, if we understand the church as a meeting or assembly for the faithful, the hearing of the spoken word, mainly of the preaching, must have been a priority in the first Christian communities (St. Paul wrote: "Faith comes from hearing the message, and the message is heard through the word of Christ.")[Romans 10:17, New International Version]. However, in a unique and characteristic way, and because of the lack of a specific and specialized architecture in the long interval of time that separates the Greek Odeon from the modern-day auditorium or music hall, the church certainly stands out among different types of buildings as the place that, for a time, witnessed the birth and performance of musical compositions. In fact, many historians believe that these buildings have played an important role in the evolution of music. The contribution of the Church of Notre Dame in Paris to the development of polyphony is a characteristic example (Abraham [4]).

Therefore, it is interesting that, upon consulting numerous documental sources on the history of architecture and the history of music, we have found almost no evaluations of or references to acoustics in churches. This may be because form has long taken precedence over function, even liturgical function, in the development of church architecture. Acoustics is one of the functional aspects that has been overlooked. We must not forget that for many centuries the history of architecture has been the history of church architecture.

In this paper, we hope to address the specifics of acoustics in churches. Also, the focus derived from the fact that there are professors from the school of architecture on our team (Sendra and Navarro are architects and Zamarreño is a physicist). This will not have an exclusively technical focus, as would be expected within the discipline known as "architectural acoustics", but will also include an architectural evaluation where history, trade and building are very much present.

Following the criteria expounded, we have divided this chapter into three sections. In the first one, we will make an acoustical analysis of the different types of churches, from the first Christian churches to the Late Baroque churches. We believe this analysis will be of great use in understanding the problem of church acoustics, and that it will prove helpful to architects who are called on to carry out a church rehabilitation project. Now we will analyse and evaluate the acoustics of a type of church that is widespread in Andalusia (southern Spain), and especially in the city of

Seville: the three-nave Mudejar–Gothic church with a timber roof (in fact, historians refer to this type of church as the “Seville parish type”). Most of these churches were built between the thirteenth and fifteenth centuries. This type of church is characterized by its acceptable acoustic conditions (Sendra et al. [5, 6]) for music and for any activity for which the spoken word is the sound source, which allows us to draw a series of conclusions. We will also describe the methodology we used for the different acoustic studies in churches, which take into account their spatial peculiarities. In the third and last section, we explain several acoustic rehabilitation projects in which our team has participated in recent years. The rehabilitation goals and the project decisions adopted to correct the deficiencies observed will be highlighted. Finally, an acoustic evaluation carried out after the interventions described will be included.

2 The acoustics of different types of churches

2.1 Early Christian and medieval churches

The early Christian church held its meetings in private homes (*domus ecclesiae*) or in existing places of an appropriate size for gatherings. The first Christian churches were built for the most part after the Milan edict in the year 313, when their religion, favoured by Emperor Constantine, could be openly proclaimed and their numbers started to grow.

As a model for their churches, understood as assemblies or gathering places for a congregation of believers, they chose the Roman civil basilica, which had a longitudinal, cross-shaped floor plan. This highly symbolic design proved ideal for their evangelistic aspirations: religious teaching, propagation of the faith through preaching, and the conversion of great numbers of people. The separation between the "church of God" associated with the clergy and the "church of the people" associated with the laymen can be seen in two fundamental parts of the basilical floor plan: the presbytery and the nave.

The structural system of the Roman, and later Christian, basilicas was lintelled, with light timber roofs transmitting small loads and thrusts. The only vaulted part was the apse. The suitable and harmonious proportions, relatively low timber roofs, and the absence of large sections of bare, reflectant walls were conducive to much better interior acoustics than we would later see in medieval Romanesque and Gothic churches.

Sometimes, the main nave had a horizontal wooden coffered ceiling, which was very appropriate for helping diffuse interior sound (Shankland and Shankland [7]). An example of this is the Santa Maria Maggiore Basilica (Figure 1), one of the few surviving examples of these early churches.

The transition from early Christian to Romanesque architecture came about slowly. Because of civil conflicts and disturbances in the turbulent medieval context, the

Figure 1. Inside St. Mary Maggiore´s Basilica.

Figure 2. Inside the San Vicente de Cardona Church.

timber roof trusses of many churches burnt, causing enormous damage. This led the Romanesque builders to design a more durable, fire-resistant roof. A structure made primarily of stone with a vaulted roof was chosen (Figure 2).

This decision greatly altered the acoustic conditions in early Christian churches. The substitution of the wooden roofs of the early Christian basilicas by hard, reflective, focalizing stone vaults in the medieval church was a great step backwards as far as their acoustics were concerned.

At any rate, in contrast with the early Christian churches, the intelligibility of the spoken word was no longer of great importance in the medieval church. The faithful, for the most part, no longer understood Latin, which was the liturgical language (Riguetti [8]). Preaching had almost disappeared, giving rise to a more natural religiosity grounded in mystery. Churches gradually discarded their role as a place for believers to gather and participate, becoming instead sacred temples for the celebration of solemn rituals officiated by their ministers (Jungmann [9]).

On the other hand, the sonority of these churches was very appropriate for the interpretation of Gregorian chants. The prolonged notes of the chant due to the high values of reverberation time produced a fullness of sound and a sensation of hearing harmonics that were not present in the melody but were created thanks to the acoustics of the large, reverberant church space.

The evolution from the dark Romanesque church to the ethereal Gothic cathedral brought even worse acoustics. The large size, above all the height of the main nave, the large reflective walls, and the vaults caused not only excess reverberation, they even produced echoes.

However, the stained glass had a beneficial effect when it covered enough area. Side chapels, introduced in the Gothic period and more fully developed in the

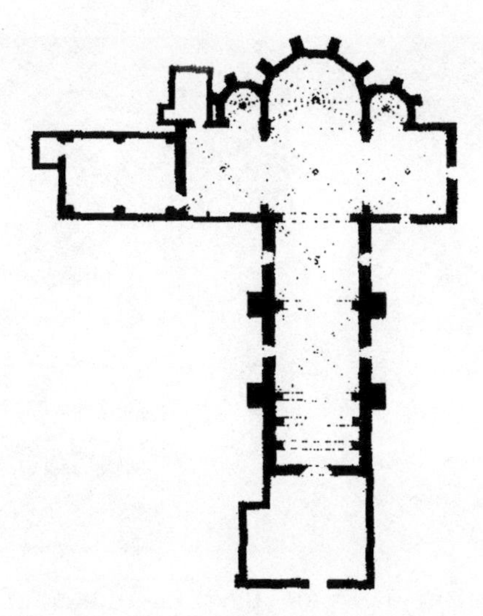

Figure 3. Plan of the Santo Domingo de Tuy Church.

Figure 4. Inside the San Isidoro Church in Seville.

Renaissance, acted as resonators, contributing to sound absorption and diffusion, especially for low-frequency sound. Also, these chapels are often ornamented, which contributes even more to absorption and diffusion.

With these exceptions, there was hardly any other sound absorption in the Gothic cathedral than that provided by the congregation, which stood, often crowded together, during the church services. There were still no seats for the congregation during this period. Pews were introduced by the Protestants because of their long church services, and the Catholic Church began to use them later (Jungmann [10]).

Toward the end of the Middle Ages in the thirteenth century, a number of mendicant orders bent on reform and purification arose in response to demands for greater spirituality in the church. The Franciscans and the Dominicans were two of the most significant of these orders.

Although the mendicant orders did not have one particular type of church building and the principles governing the construction of their temples depended greatly on the building traditions of the place, their spirit of reform led them to design their churches with two main goals in mind, liturgy and preaching. The Dominicans set a limit to the size of their churches, especially the height of the nave. For the Franciscans, the only part of the church that should be vaulted was the apse. The naves of their churches usually had a timber roof resting on transverse arches, although in many cases these roofs were replaced by vaults. Also, the Franciscans chose the single-nave model (Figure 3), rather than the triple-nave Cistercian model. This open floor plan made it easier for the congregation to see and hear the preacher.

These orders, who preached in the common language instead of Latin to be better understood by the congregation, brought a renaissance in preaching.

Figure 5. Inside the St. Lorenzo of Florence Church.

Figure 6. Inside the St. Andrea of Mantua Church.

Mudejar churches were also built in the Spanish Middle Ages, reaching their greatest splendor in the thirteenth and fourteenth centuries. Mudejar churches usually had a vaulted presbytery as a result of the stylistic evolution from Romanesque to Gothic, and the naves had timber roof trusses following Moorish tendencies (Figure 4). Many of these roofs have been lost to fire.

The Mudejar building tradition was very widespread in Andalusia and was used mainly in the parish churches. Their small size and, above all, use of timber roof trusses rather than vaults to cover the naves are the main causes of their generally suitable acoustic behaviour in comparison with their Gothic contemporaries (Sendra and Zamarreño [11]).

2.2 Renaissance and Tridentine churches

Tuscany was the artistic centre of the Quattrocento, and Brunelleschi and Alberti were its two great architects. They developed a church type with a central floor plan and another church type with a longitudinal floor plan. These two types served as models for Renaissance religious architecture.

In the San Lorenzo and Santo Spirito churches, Brunelleschi recovered the type of basilical church used by the early Christians, although it was adapted to the needs of the moment. The main innovation was the introduction of side chapels. The naves of both churches have coffered wooden ceilings, the side chapels are vaulted, and the crossing has a dome (Figure 5).

From the point of view of acoustics, the triumph of horizontality over the ever-increasing verticality of the Gothic, more harmonious proportions, a coffered wooden ceiling for the main nave, and the absence of large reflecting wall surfaces could only have a beneficial effect.

Figure 7. Inside the San Francesco (San Salvatore) al Monte Church (Il Cronaca).

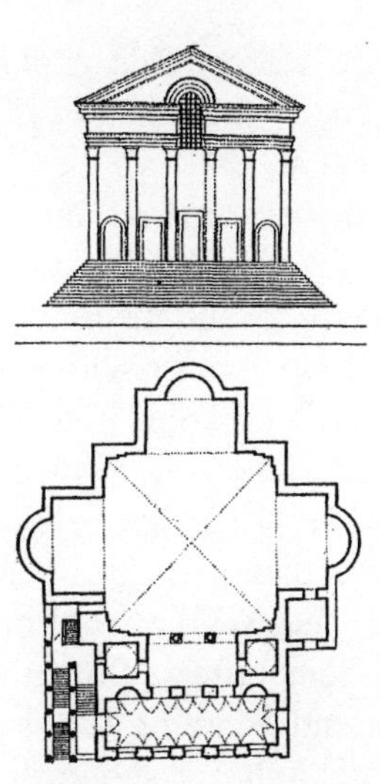

Figure 8. Plan and section of the St. Sebastian of Mantua Church (Alberti).

Alberti, however, in what would be the first treatise on Renaissance architecture, *De re aedificatoria,* specifically in Book VII, voices the opinion that, in contrast with the basilica, churches should have vaulted roofs to provide them with greater dignity and guarantee of permanence. San Andrea in Mantua, covered with a grand coffered barrel vault (Figure 6) is his prototype of the longitudinal church.

This does not mean that Alberti ignored the beneficial effects of timber roofs and the negative effects of vaults on acoustics, other sections of his treatise indicate that he was familiar with them (Benedetti [12]), however, in his opinion and in the opinion of his generation, church acoustics were not a priority.

The substantial change in consensus about church acoustics occurred in the second half of the sixteenth century and was influenced by decisions of the Council of Trent and by the importance the council lay on preaching as an instrument of the Counter-Reformation.

This attitude already had precedents in the architecture promoted by the mendicant orders starting in the thirteenth century and in the architecture of certain reform movements within the Church during the first half of the sixteenth century. These reform movements, classicistic preoccupations aside, advocated an architecture for all mankind (Benedetti [13]) in which simplicity, common sense and reason would predominate (Figure 7) rather than the court (imperial) architecture promoted by Renaissance treatises (Figure 8).

The intelligibility of text and even of music was fundamental for the Catholic reform, and was advocated by the council. There is documentary evidence during this period of time of concern for acoustics in the construction of the churches of some religious orders, especially those of the Jesuits and Franciscans, to promote the homiletic effort. This explains their preference for the single-nave church with a timber roof, which had been previously used with success by some mendicant orders (Sendra et al. [14]).

We describe below four important documents from this period of reform that bear witness to how church roofs are related to interior acoustics. The earliest example is the Franciscan and the remaining three are Jesuit.

The Franciscan document has advice given by Francesco Giorgi ten years before the Council of Trent in a 1535 memorandum to the builders of San Francesco della Vigna, designed in Venice by Sansovino. From it, we quote the following lines (Wittkober [15]) that refer to Giorgi's acoustic assessment of the church and his recommendations in this sense: "I recommend to have all the chapels and the choir vaulted, because the word or song of the minister echoes better from the vault than it would from rafters. But in the nave of the church, where there will be sermons, I recommend a ceiling (so that the voice of the preacher may not escape nor re-echo from the vaults). I should like to have it coffered with as many squares as possible, with their appropriate measurements and proportions (...) And these coffers, I recommend, amongst other reasons, because they will be very convenient for preaching; this the experts know and experience will prove it."

In reality, Giorgi was distinguishing in the church what, five centuries later, the English physicist Bagenal would call "cave-like acoustics". These acoustics were more appropriate for musical performances and hymns, while open-air acoustics were more appropriate for speech (Forsyth [16]).

This should not surprise us coming from a religious order such as the Franciscans, concerned with the quality of preaching since their foundation, long before Counter-Reformation ideals came into vogue; and especially in Venice, where music in general and religious music in particular were to develop so brilliantly. Giorgi's theory that the coffered ceiling would favour preaching reveals a certain knowledge of acoustics and, specifically, of the principles of sound diffusion.

The coffered ceiling that Giorgi proposed for the nave was never built, for reasons unknown to us.

The other three documents mentioned are from the Jesuits, a newly-founded religious order at that time that played an active role in the council and that placed special emphasis on church acoustics.

One of these three documents refers to a Spanish church: the first Jesuit church built in Madrid, inaugurated in 1557 by Philip II, designed by Bartolomé de Bustamante and, unfortunately, no longer in existence. In this document, dated in 1569 and replicated by Rodríguez [17], the provincial head in Toledo wrote a letter to the general head of the Jesuits in Rome praising the excellent acoustics of the church mentioned above, in spite of its large size (38 m long by 11 m wide), and attributing this quality to the timber roof. He ends the letter expressing his disappointment at the fact that the churches are not always roofed in this fashion and commenting that this system would save money and also improve the intelligibility of the sermons, whereas vaulted roofs were the cause of poor acoustics.

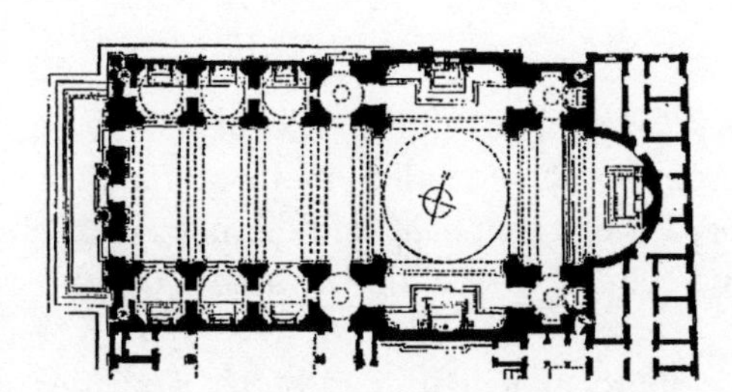

Figure 9. Plan of the Il Gesù Church in Rome.

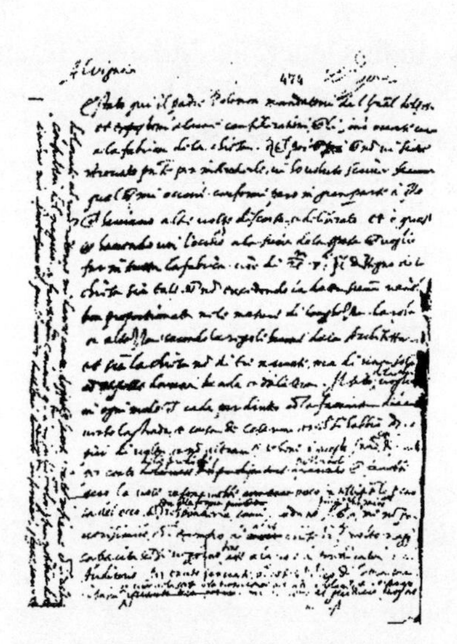

Figure 10. Cardinale Alesandro Farnese to Giacomo Barozzi dello il Vignola.

The second of the Jesuit documents refers to the mother church in Rome, Il Gesù (Figure 9), the great Counter-Reformation temple designed by Vignola. This church was built under the initiative of Cardinal Farnesio. The Jesuits, and more specifically Father Tristano, St. Francisco de Borja's trusted assistant and the man who filled the *consiliarius aedificatorium* post of the Society of Jesus until his death, pressured the cardenal, to no avail, to build a flat timber roof instead of a barrel vault (as was finally built) precisely for acoustical reasons.

The document alluded to is a letter from Cardinal Farnesio to Vignola dated in August 1568 (Figure 10). In it, we can see the differences in criteria between the cardinal and the Jesuits about the construction of Il Gesù and the Jesuits (Pirri [18]). Specifically, the cardinal informs Vignola that the church is to have only one nave instead of three with chapels on either side and a vaulted roof, although the Jesuits thought that would hinder preaching. "They think that the voice will resound unintelligibly because of the echo ... (more than with the flat wooden ceiling) ... although this does not appear probable to me because of the example of what occurs with other larger vaulted churches, which adapt well to the voice of the preachers." We do not know which churches he was referring to.

Tristano's preference for timber roofs, for acoustical reasons among others, appears in documents referring to another one of the key Jesuit Counter-Reformation churches related to the figure of Cardinal Borromeo: the church of St. Fidele of Milan, designed by Tibaldi. Father Pirri [19] narrates what occurred with this church's plan and roof: "The flat roof was a regulation that answers to Tristano's personal preferences. We have proof in a letter from Leonetto Chiavone, rector of the Jesuit residence in Milan, to Benedeto Palmio about Pellegrino Tibaldi's design for the St. Fedele Church in that city. In this letter, the Milanese Jesuits were wondering what would happen to the plan in Rome because they were familiar with Tristano's prejudices concerning vaulted structures with a curved apse."

As the letter explains, the reason for this concern was the poor intelligibility due to the echo that could be produced. Finally, this church was not built with a wooden roof either.

2.3 Baroque and Late Baroque churches

From the second half of the sixteenth century, when the reformist ideals of some religious orders defending a modest, simple architecture prevailed, there was a clear transformation toward what could be considered the triumphant embodiment of the Catholic Counter-Reformation. The Baroque church was the symbol of this triumphant spirit.

From the acoustical point of view, Baroque churches usually had better acoustics than their classical predecessors, mainly because of their ornamentation. This effect can be noticed mainly with mid-range and high-pitched sounds. Mouldings, pilasters, entablatures, cornices, columns and capitals adorned with ova, garlands and beads help diffuse sound. The same occurs with carved stucco, wooden appliqués, altarpieces, organs, inner doors and other furniture. On the other hand, the abundance of side chapels in this type of plan diffuses more low sounds.

For many people, the oval floor plan is the clearest example of the Baroque. This form already had precedents in the second half of the sixteenth century and in the early seventeenth century as an attempt to combine classical aspirations of churches with a central floor plan with the liturgical needs set forth in the Council of Trent that required a longitudinal floor plan.

Even the Jesuits themselves had proposed oval floor plans for some church buildings, among other reasons, because of acoustical considerations (Vallery-Radot [20]). Although we do not know how they justified this dubious assertion, the oval did become the favourite shape for opera theatres because its focalizing effect was useful.

It was in the early seventeenth century, specifically in 1604, when Kepler defined the ellipse as we now understand it. Right away, a number of papers and studies on the problems of light and sound reflection were written. One of the most outstanding was *Phonurgia Nova*, published in 1673 by a Jesuit Scholar named Kircher, who gave us a unique graphical representation showing that any sound ray with its origin at the focus of the ellipse (in this case, an ellipsoid vault), after reflecting off a surface, passes through the other focus (Figure 11).

However, the oval did not prove to be an appropriate form for large churches, nor did it permit much variation. It was not until the appearance of Baroque architects Borromini and Guarini that new modes of expression were found. The dynamism of the architectural spaces that they designed led to the use of alternate concave and convex forms, which could only have beneficial effects for the acoustic conditions of the churches (Figure 12).

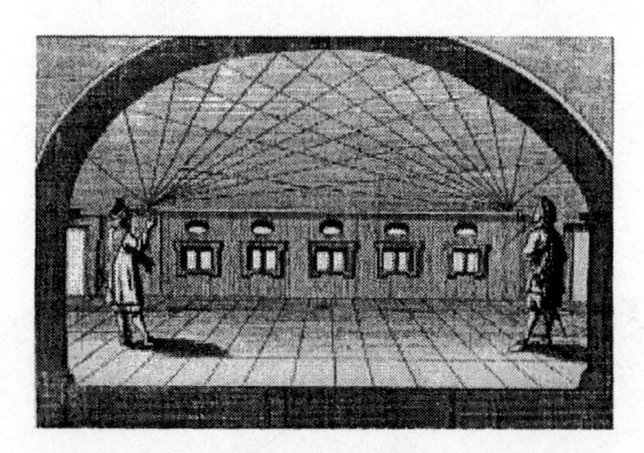

Figure 11. Focal points of the oval. Phonurgia Nova.

Figure 12. Inside the Sant'Ivo Church in Rome (Borromini).

Figure 13. J. Berain. Funeral decorations for Prince Condé in Notre-Dame in Paris, 1687.

The influence of these two architects was enormous in Late Baroque Central European architecture toward the end of the seventeenth century and the first half of the eighteenth century, scene of outstanding musical performances of the best composers of the period. Still today, these churches are considered the ideal venue from the optical and acoustic point of view for the performance of Baroque music.

We do not want to end this section without mentioning one of the Baroque operations with the greatest repercussions on church acoustics: ornamentation with large tapestries and hangings (Figure 13), especially for the main civil or religious celebrations of joyful or sorrowful occasions. Beatifications, canonizations, exequies and funerals, proclamations of dogmas and processions led to a rich ephemeral architecture of which many written and iconographic testimonies remain.

This will be a constant in the coming years: the ephemeral modification of the absorption properties of the hangings to adapt them to musical and choral interpretations (Schnoebelen [21]). The abundance of tapestries and hangings, velvets and damasks, together with the profuse general decoration and the large number of people dressed in heavy, flowing clothes could only increase sound absorption in these churches, damping echoes and high-range reverberations.

3 The sound field in Mudejar–Gothic churches with timber roofs

3.1 Mudejar churches in Seville

When the Christians recovered southern Spain from the Arabs in the thirteenth century, they decided to build their temples over the mosques, thus symbolizing the new authority. For their buildings, they decided to follow the so-called "proto-Gothic", which at that time was the highest form of artistic expression in the Christian western world. We know that Fernando III, and later Alfonso X, had among his craftsmen stonecutters from the north who were experts in the new architecture in vogue in Europe.

The effort of successive Castilian monarchs to maintain cities such as Córdoba or Seville that often surpassed those of their kingdoms in size and importance caused the cities to be partitioned according to the three linguistic, religious and cultural groups: the Christians, the most important group; the Jews, fewer in number and located in a kind of separate city; and the Moors, the smallest colony because most of the Arabs had emigrated to Islamic kingdoms.

The walled enclosures were restored and even enlarged, and the fortresses were used as palaces for the monarchs. The dwellings were readapted, joining houses to provide new residences. Many mosques were repaired and adapted for the new religion, changing their layout so they faced the proper direction and using their minarets as bell towers. There immediately arose an artistic confrontation between the Gothic Castilian culture and the Muslim Andalusian culture; this conflict gave birth to the most distinctive artistic movement in medieval Spain: the Mudejar style.

The cultural assimilation of the conquered Moors and the fascination of the Christians with the monuments they found in the conquered cities, in some senses thought superior, of *Al Andalus* and close ties with the unconquered territories are factors that help explain this unique artistic phenomenon. The Mudejar style cannot be understood as Islamic art or as Western Christian art, but rather as a link between the two; it is a uniquely Spanish artistic movement.

Figure 14. Some morphological characteristics of Mudejar churches: Above, the most common apsidal forms. In the centre, some plans with a collar beam roof. Below, a typical section of a Mudejar church.

The first buildings were made entirely by Christian stonecutters, but this situation could not continue for very long because the increasing number of churches to be built required the help of Moorish masons. These masons contributed their ideas and construction methods little by little until they fused with those of the conquerors.

The vast amount of work to be done together with the lack of suitable stone in the area led to the use of brick masonry (in which the Moorish masons were specialists) as the normal building material. Stone was reserved for the most important parts: portals and apses. In reality, the Moorish construction system began to take precedence over French stonework because throughout the territory the building trades were well organized.

The Mudejar churches in Seville follow the older Cordovan model. Morphologically, they are characterized by a stylistic dualism (Figure 14): a vaulted

Figure 15. Tower, portal and interior of Santa Marina Church in Seville.

Gothic apse and a body of three naves with a timber roof (collar beam in the main nave) of Moorish origin. Its brick walls are complemented with portals and a stone apse. The supports are also clearly Islamic, with quadrangular or sometimes octagonal pillars, with raised brick mouldings as decoration.

Pointed arches, round arches or segmental arches rest on these supports. In some churches, such as San Marcos, horseshoe arches with a surround are used. The narrow, elongated, Gothic voids of the Mudejar churches, with a pointed arch and mullion, are normally used in polygonal apses. The naves are illuminated almost exclusively with rose windows and oculi located at the foot of the nave.

The portals project from the face: the earlier ones were pointed or horseshoe arches with a surround, and the later ones are splayed arches decorated with large bead mouldings, capitals, sculptures, sebka panels and small roofs. The towers are almost always a pre-existing Muslim minaret with an addition and a belfry. Many of the towers are post-Islamic, but they show the stylistic influence of the minarets (Figure 15).

Among other elements of particular interest, we find the funeral chapels that have been added successively to the side naves, donated by wealthy families, and which, on some occasions, are housed in remaining sections of pre-existing mosques. These chapels usually have a square floor plan with octagonal domes supported by pendentives, imitating the Al Andalus qubba, a design that was introduced around the year 1400.

According to Angulo [22], the first Mudejar church built in Seville was Santa Marina; this church established the so-called "Seville Mudejar" type. It would be the first of a series of temples built during the reigns of Fernando III and Alfonso X including Santa Marina itself, San Julián, Santa Lucía and Santa Catalina. Later churches include San Lorenzo, San Vicente, San Isidoro and San Marcos. Later, in the year 1356 during the reign of Pedro I, there was a severe earthquake in Seville that seriously damaged many of these temples. Pedro I initiated an important restoration

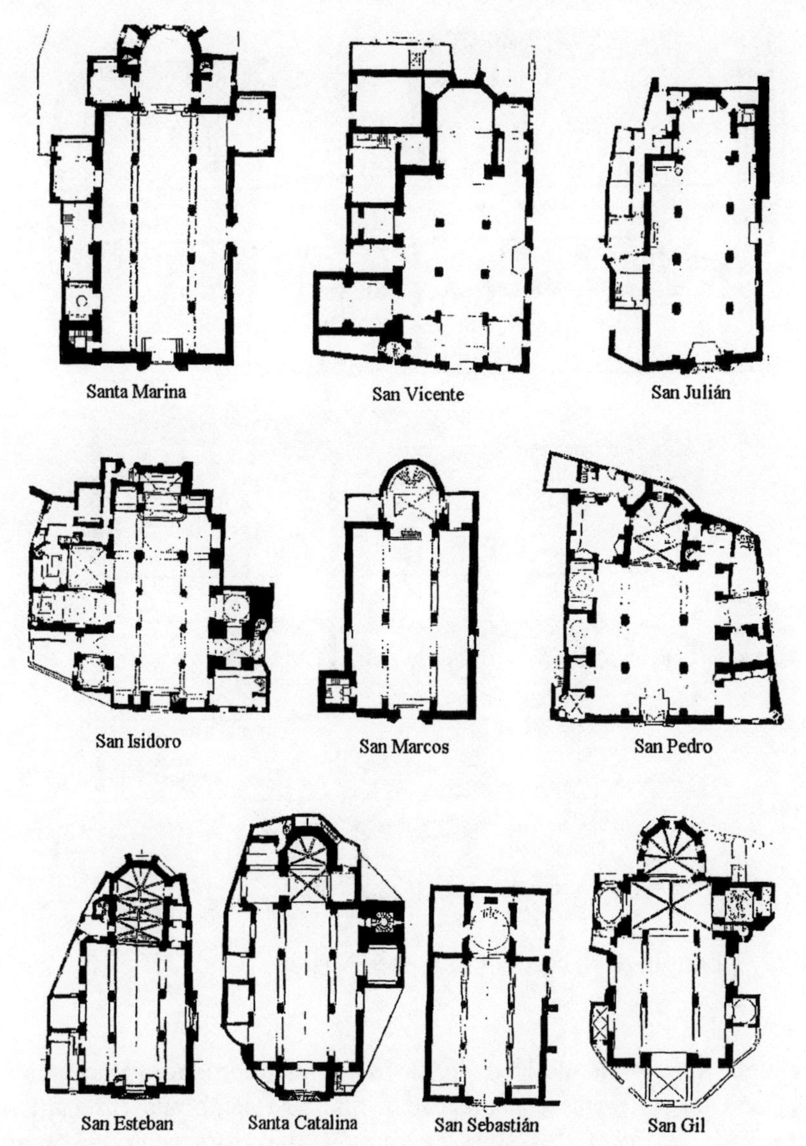

Figure 16. Plans of the ten Mudejar Gothic churches analysed.

program and built a number of new churches. The churches built have been referred to by Angulo as "The Group of 1356", and include Omnium Sanctorum, San Esteban, San Andrés, San Roque and San Pedro Churches.

As can be seen in Figures 16, 17 and 18, which show the plans of the ten churches studied with their longitudinal and cross-sections, a completely original church type was established and adhered to in the construction of each of these ten churches with only slight variations and modifications of scale.

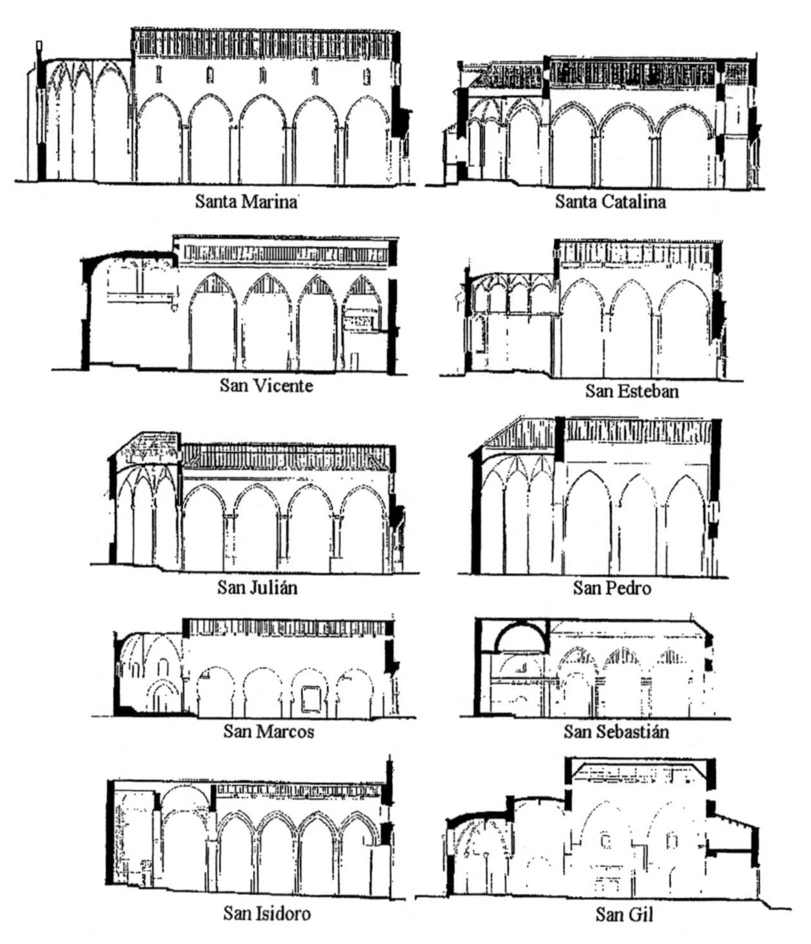

Figure 17. Longitudinal sections of the ten Mudejar–Gothic churches analysed.

The church type thus defined in Seville was reproduced throughout lower Andalusia, and its style was consolidated in the fourteenth and fifteenth centuries. During the early years of this style's evolution, the city's central monument, the cathedral, was built. This edifice would stamp all successive Andalusian construction with a new character and even influence the Mudejar design, which was then emerging in the Kingdom of Granada, still in Muslim hands.

To understand the Mudejar style, it is important to know that it had its origin in each reconquered region at different times and was coeval with various types of artistic expression found in each place, including the type of Mudejar design that had previously developed in the border zones. Therefore, each region's precedents are different, conferring the style with great diversity. Also, as we mentioned before, it did not appear immediately after the reconquest of the territory, but arose after a period of time in which the conquerors made an effort to give it the first Christian imprint. On the other hand, we should bear in mind that the Spanish Mudejar, on the whole,

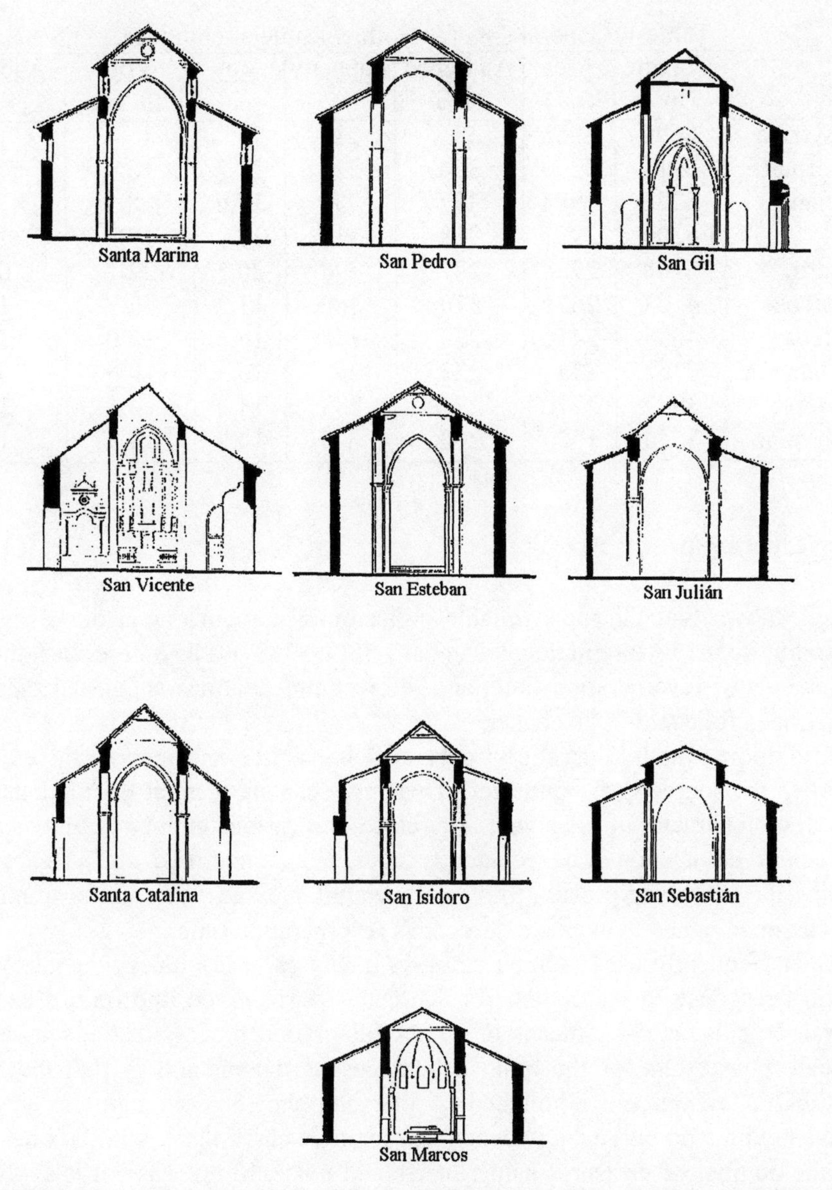

Figure 18. Cross-sections of the ten Mudejar–Gothic churches analysed.

survived a long time, outlasting the western Gothic or Romanesque styles, and the successive Islamic Almoravid, Almohad and Nasrid styles with which it coexisted (Borrás and Sureda [23]).

Table 1 summarizes some general data of interest for understanding the acoustic behaviour of the ten churches analysed.

Table 1. General data for Gothic–Mudejar churches.

Church	Volume (m^3)	Area (m^2)	Audience area	Capacity	Vol. / seat	Vol. / aud. area	Aud. area / seat
Santa Marina	8696	3988	531	530	16.4	16.4	1.0
San Vicente	6915	3290	366	327	21.2	18.9	1.1
San Julián	6226	2984	337	354	17.6	18.5	1.0
San Gil	6200	2931	318	328	18.9	19.5	1.0
San Pedro	6108	2758	277	298	20.5	22.1	0.9
San Marcos	4763	2671	370	360	13.2	12.9	1.0
San Esteban	4683	2416	275	278	16.8	17.0	1.0
Santa Catalina	4362	2251	223	267	16.3	19.6	0.8
San Isidoro	3947	2270	277	283	14.0	14.3	1.0
San Sebastián	3550	1980	262	240	14.8	13.5	1.1

3.2 Reverberation

It is well known that the most reliable indicator of a room's acoustic behaviour is reverberaton time *T*. International standard ISO 3382 (1985) describes the field measurements of reverberation time, and the recommendations suggested there have generally been followed in this paper.

The traditional method uses a broadband noise source and records the extinction curve when the source is disconnected. This curve can show great fluctuations due to the stochastic character of the signal. Therefore, the various extinction curves at each measurement point should be coherently averaged (Zamarreño and Algaba [24]). Extending the averaging process to obtain a spatial average, more uniform extinction curves can be obtained to evaluate the room's reverberation time.

Another frequently used method employs a shot as an excitatory signal to obtain the room's response to the signal. As Schröeder [25] proved, the room's extinction curve can be calculated by means of a back integration process of the square of the impulse response. One of the main advantages of the method is that the signal's stochastic fluctuations are eliminated so that one shot at each point is enough to provide an extinction curve equivalent to the average curve that would be obtained by an infinite number of measures using interrupted noise. In this case, it is also a good idea to calculate the spatial average we mentioned earlier.

The impulse response was the method used in this study. In addition to the advantages we just pointed out, this method is also a simple way of obtaining the room's other acoustic parameters related to the early to late energy relations, such as clarity or definition.

The shot was fired with a 9 mm starting pistol at the place where the natural source would usually be located in the presbytery by the altar. Two shots were fired at each measuring position to ensure that at least one of the recordings would be usable. One position is used for the shot, and between four and seven microphone positions are

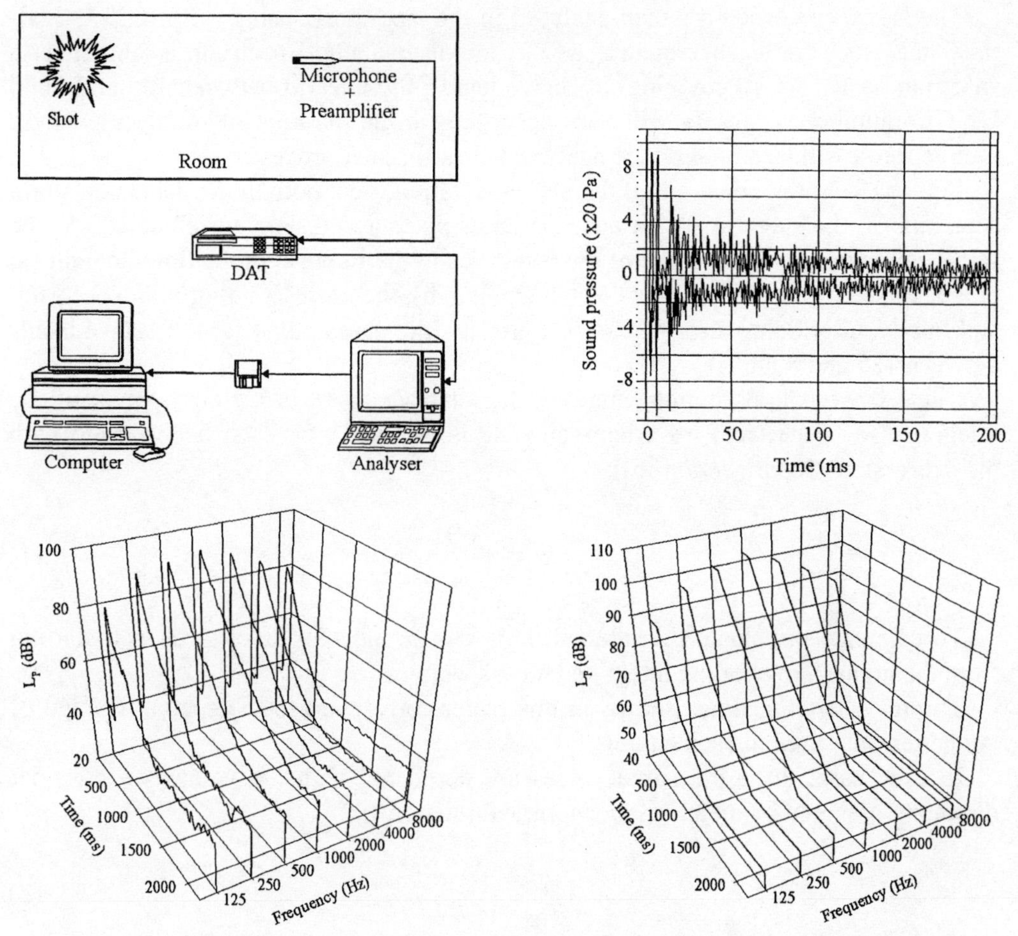

Figure 19. Reverberation tests: measurement system, response to the impulse in the time domain, multispectrum of the response to the impulse, integrated multispectrum to determine the extinction curves.

used for the receiver, covering both the central naves and the aisles. The microphone is situated at approximately the height of the head of a person seated in one of the pews with which the church is furnished (approximately 1.20 m from the floor).

The response to the impulse at the different points chosen inside the church was recorded with a microphone (B&K 4165) with its respective preamplifier (B&K 2639) connected through a microphone polarization source (B&K 2804) to a Sony DAT PC204 magnetic recorder. The setup of the DAT input was adjusted in each church to take advantage of its dynamic range, avoiding overload.

In order to find the distance, in each case, between the emission point and the reception point, one of the DAT channels recorded an electric impulse synchronized with the shot. We can find this distance by measuring the time delay between reception of the electric impulse and reception of the direct sound.

The signals recorded are then analysed in the laboratory, using a B&K 2133 real-time analyser. For each response to the impulse, a multispectrum is obtained (a spectrum each 1/64 s), covering, in octave bands, the interval between 125 and 8000 Hz. The number of spectra will vary according to the duration of the decay. These multi spectra will be stored on the analyser's disk for later processing.

Figure 19 shows a diagram of the instrumentation used, both in the data acquisition stage and in the stage in which analysis takes place and results are obtained. At the same time, we can observe one of the responses to the impulse in the time domain (in one of the recordings from Santa Catalina Church), the corresponding multispectrum, and finally, after Schröeder's integration process, the decay curve for the octave bands between 125 and 8000 Hz.

Using the functions implemented by the analyser itself, based on slope m of the integrated multispectrum, reverberation time is evaluated for each octave band with the expression (Sendra *et al.* [26]):

$$T = -\frac{60}{m} \tag{1}$$

To calculate the slope, the interval between –5 and –30 dB was chosen starting from the highest level of the integrated multispectrum.

The reverberation times shown in this paper correspond to the spatial average of the points measured in each church.

In each case, these measured values are compared with the values we consider optimum, obtained by means of the empirical relation:

$$T_{\mathrm{OP}} = k\, u\, i\, V^{1/3} \tag{2}$$

In this expression, coefficient k depends on the frequency, coefficient u is derived from the use of the room, and coefficient i takes into account the existence or lack of public-address system (PA). The values for these parameters are shown in Table 2.

Table 2. Obtaining the optimum reverberation time.

f	125	250	500	1000	2000	4000
k	1.30	1.15	1.00	0.90	0.90	0.90
u	Speech: 0.75			Music: $0.08 < u < 0.1$		
i	with PA system: 0.85		without PA system: 1			

Since the tests were carried out while the churches were empty and the furniture absorbs very little sound, we have estimated the expected times in relation to the degree of occupation. Therefore, for our first estimate, we used Sabine's equation, the absorption values for the empty room obtained from the reverberation times measured, and the absorption per person for the estimated audience.

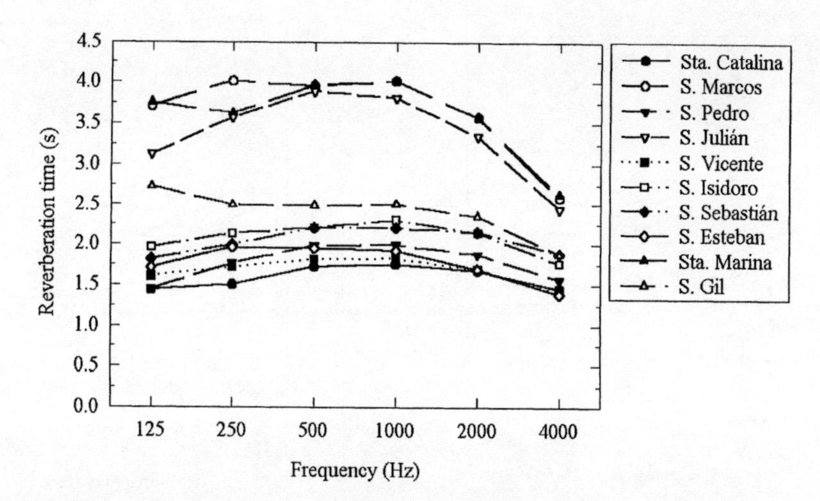

Figure 20. Reverberation time in octave bands for the ten churches.

Since in this type of building it is normal to find connected spaces (side chapels, presbytery), we used the Knudsen method [27] when necessary to perform the analysis. Specifically:

$$\text{If} \quad p \leq 2h \;\Rightarrow\; T = 0.161\frac{V_1 + V_2}{A_1 + A_2} \tag{3}$$

In other words, the two spaces (with volumes V_1 and V_2 and absorptions A_1 and A_2) are considered one room, where p is the depth of the smaller space and h is the height of the opening between them. In case this condition is not fulfilled:

$$\text{If} \quad p > 2h \;\Rightarrow\; T_1 = 0.161\frac{V_1}{A_1 + A'}, \quad T_2 = 0.161\frac{V_2}{A_2 + A'} \tag{4}$$

where A' is the virtual absorption of the surface separating the spaces, calculated from the absorption coefficients in Table 3.

Table 3. Absorption coefficients for the virtual surface of connected rooms.

p/*h* ratio	125 Hz	500 Hz	2000 Hz
2.5	0.30	0.50	0.60
3.0	0.40	0.65	0.75

The results of the tests for each of the churches are listed in Table 4 and are shown as a graph in Figure 20. Figure 21 compares the estimated values with the optimum and measured values for two of the churches analysed: Santa Marina, which is large

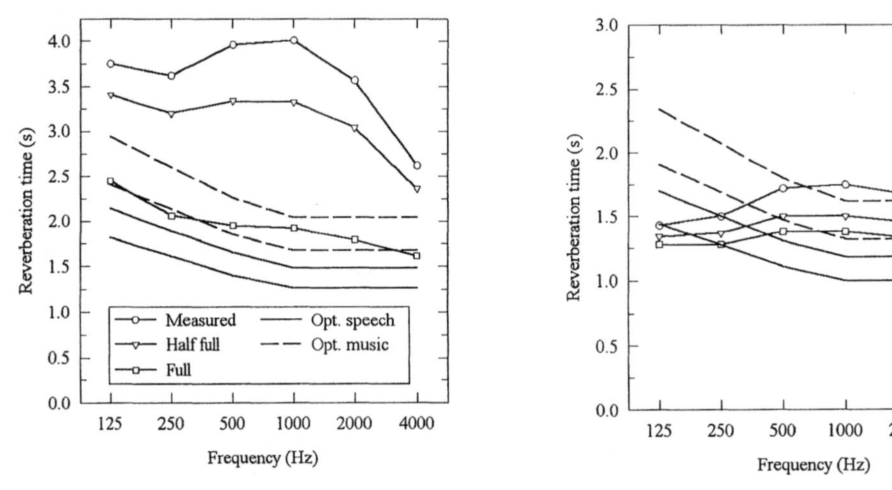

Figure 21. Measured and estimated values for one of the reverberant churches (Santa Marina) and one of the dead churches (Santa Catalina).

and reverberant, and Santa Catalina, which is small and dead. The optimum band for speech is bounded by two polygonal lines plotted from the optimum values of the reverberation times with or without the PA system, while the optimum band for music was determined according to the appropriate values for chamber music, on the one hand, and for liturgical music, on the other. The half-full option means that only the wooden pews were considered full. The full option takes into account a standing audience density of one person per square metre where there are no pews.

Table 4. Measured reverberation times for Gothic–Mudejar churches.

Church	125	250	500	1k	2k	4k
Santa Marina	3.75	3.62	3.96	4.01	3.57	2.62
San Vicente	1.60	1.71	1.81	1.82	1.68	1.40
San Julián	3.12	3.57	3.89	3.81	3.34	2.44
San Gil	2.72	2.49	2.48	2.50	2.35	1.87
San Pedro	1.45	1.76	1.98	1.99	1.87	1.56
San Marcos	3.71	4.01	3.95	4.01	3.57	2.58
San Esteban	1.71	1.96	1.95	1.92	1.69	1.38
Santa Catalina	1.43	1.50	1.72	1.75	1.67	1.45
San Isidoro	1.69	2.13	2.21	2.30	2.14	1.76
San Sebastián	1.81	1.99	2.20	2.20	2.15	1.88

Figure 20 shows the existence of two groups, one of which has higher reverberation times, formed by Santa Marina, San Marcos and San Julián churches, and which have suitable reverberation values for liturgical music only when they are completely full, but are too reverberant for speech. San Gil church marks the transition to the second group, in which close to optimum values are achieved for verbal

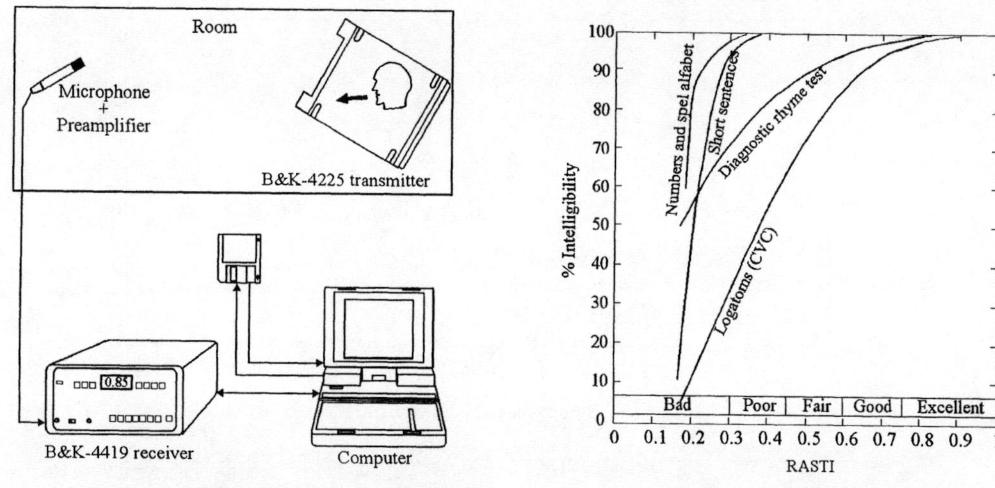

Figure 22. Intelligibility test: diagram of the measurement system and of the relation between the RASTI index and several subjective tests.

communication even with a smaller audience present. In all cases, reverberation times for low frequencies are, paradoxically, less than at medium frequencies. This is attributed mainly to the timber roofs. The values for the frequencies measured depend on the degree of ornamentation in the church and the presence or absence of side chapels.

3.3 Intelligibility

From the functional point of view, speech intelligibility is one of the most important acoustical aspects to be evaluated in the rooms where oral communication plays an important role, as it does in the case at hand. Of the available subjective and objective methods for the quantitative and qualitative evaluation of speech (Sendra et al. [28]), we have chosen the RASTI (Rapid Speech Transmission Index), which is a simplified version of the STI (Speech Transmission Index), and whose main advantages are objectivity, easy utilization and repeatability. Through the use of computer simulation programs, this index can also be employed to make predictions.

Figure 22 shows a diagram of the measurement system used. We can use the nomogram in this figure to evaluate intelligibility qualitatively as a function of the RASTI index value and to establish the relationship between the corresponding values of this index and the subjective intelligibility measured by different types of tests.

With this method, a transmitter (B&K 4225) sends a signal of filtered noise to the transmission channel. This noise is within the 500 and 2000 Hz octave bands, essential for intelligibility of speech, and the signal of filtered noise is modulated in each band with low-frequency signals that try to reproduce the modulations of the human voice when speaking (1.02, 2.03, 4.07 and 8.16 Hz in the 500 Hz band, and 0.73, 1.45, 2.90,

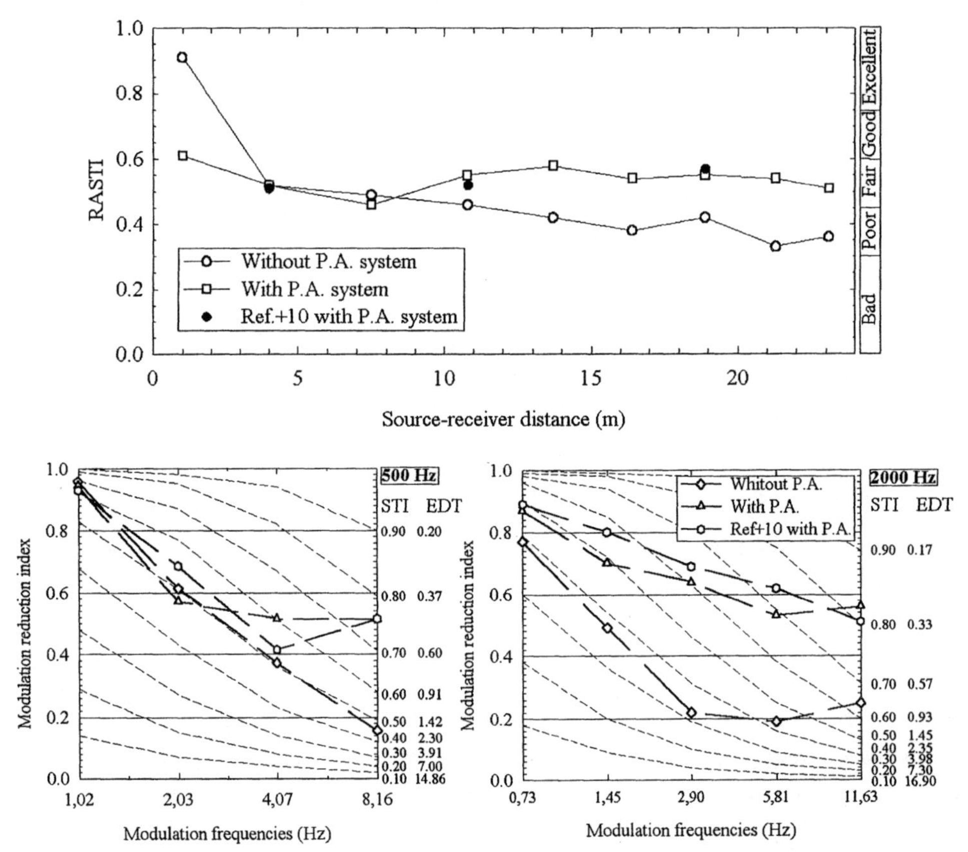

Figure 23. Variations of the RASTI index over distance in the central nave of the San Gil church and MTF for a significant point in this church.

5.81 and 11.63 in the 2000 Hz band). The transmitter's directionality is similar to that of the human head.

A microphone receives the signal and sends it to the receiver (B&K 4419). The receiver analyses modulation reduction between the emission and reception points, obtaining what has been called the Modulation Transfer Function (MTF) between the two points. This function is a measure of signal degradation between the system's input (emission point) and the output (reception point). The RASTI index is based on this measurement (Hougast and Steeneken [29]).

All of the process, including the configuration of the receiver and the storage of data on a diskette for later analysis and processing, was controlled by a laptop computer with an RS-232 interface (Figure 22). The microphone, which was set at about 1.20 m from the floor, was moved from one reception point to another in the audience area of the central nave and the aisles (an average of 13 points per church).

In all churches, the measurement period for each position was set at 32 s (the greatest possible) to minimize random fluctuations in the RASTI index caused by irregular background noise. Measurements were taken at least twice at each point to ensure that one of them could be considered correct.

Because of the room's characteristics and use, the orator will probably raise his voice when speaking to the audience, so the emission reference level was adjusted to Ref. +10 dB. This would give a reading of approximately 69 dB in the 500 Hz band and 59 dB in the 2kHz band at a distance of 1 m from the transmitter.

When possible, the test was carried out using the PA system, including the microphone, which was located about 30 cm from the transmitter's loudspeaker. A person speaking through a PA system does not speak as loudly, so for these tests, the transmitter's was adjusted to the Ref. level (59 dB in the 500 Hz band and 50 dB in the 2000 Hz band at 1 m from the loudspeaker).

Table 5 shows a summary of the average values obtained in the churches under the different measurement conditions. In some cases, the test was replicated with the PA system and with the receiver adjusted to the Ref. +10 level. The results (Table 5 and Figure 23) are very enlightening: increasing the acoustic level (10 dB in this case) and consequently the signal to noise ratio does not necessarily lead to an increase in intelligibility.

Table 5. Summary of the intelligibility test results for Gothic–Mudejar churches

Church (measurement points)	Without PA (Ref.+10 dB)	With PA (Ref.)	With PA (Ref.+10 dB)
Santa Marina (16)	0.35	0.55	–
San Vicente (14)	0.50	–	–
San Julián (13)	0.35	0.50	0.5
San Gil (14)	0.45	0.60	0.6
San Pedro (12)	0.50	0.45	–
San Marcos (13)	0.30	0.45	–
San Esteban (7)	0.40	0.50	–
Santa Catalina (14)	0.45	0.60	–
San Isidoro (15)	0.35	0.45	–
San Sebastián (11)	0.40	0.45	–

One of the aspects analysed was the difference in intelligibility because of distance, especially in the central nave. Figure 23 shows the RASTI index for the different testing conditions as a function of the distance for points situated in the central nave of San Gil Church. Throughout most of this nave, the index value rates this intelligibility as "poor" when the public-address system is not being used. The use of the speaker system causes a considerable improvement in the areas furthest away, increasing the rating to "fair". Finally, as we pointed out before, the increase in emission level (+10 dB) does not entail any additional improvement of the RASTI index.

MTF analysis is useful for finding the cause of this deterioration in intelligibility. Figure 23 also shows these functions for both octave bands for a specific point in the same church toward the back third of the main nave (19 m from the transmitter). We can observe that the presence of a PA system modifies the MTF: from the typical form

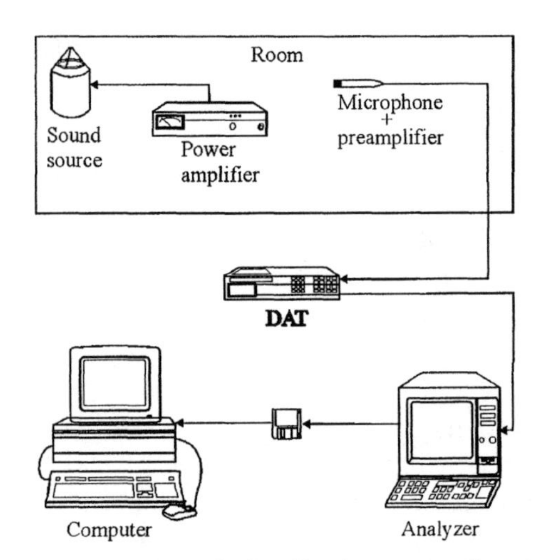

Figure 24. Sound distribution test: diagram of the measurement system.

due only to reverberation interference (dotted lines) to that due only to background noise (horizontal lines). In this case, the increase in the level of the transmitter does not result in significant modifications of the MTF.

3.4 Sound distribution

The sound distribution measurements allow us to characterize the degree of uniformity of the sound field on the inside of the room. The test consists in determining sound pressure level L_P at several points in the area of interest while a sound source placed in the most probable location of the source during everyday use (the altar) emits white or pink broadband noise.

The noise level is generated by a calibrated reference sound source (B&K 4205). The output power level was adjusted in each church to values between 95 and 99 dB (ref. 1 pW), in an attempt to reach the highest value possible without overloading the source amplifier. In this way, the level of the signal at all the measuring points and for all the octave bands of interest allows us to disregard the background noise. The instrumentation and measuring procedure used in this test are shown graphically in Figure 24.

Before the measurements are taken, and once the measure range of the DAT has been adjusted, a calibration signal of 93.8 dB at 1000 Hz is recorded. This signal is later used in the laboratory to calibrate the analysis system with a B&K 2133 analyser for processing the signal and a personal computer for subsequent data processing.

For each point (13 on average for each church), the B&K 4165 microphone was placed at a height of 1.20 m from the floor and recorded the signal generated by the sound source for 30 s. This signal was later analysed in the laboratory in octave bands between 125 and 8000 Hz, selecting a 15 s interval in the middle of each recording to perform a linear average. Also, the analyser provides the overall linear and A-weighted broadband levels.

Since the source is calibrated, knowing the value of the acoustic power emitted has allowed us to study the diffuse character of the acoustic field in each case. Therefore, we compare the levels of the reverberant field depending on the distance with those foreseen by the classical model and the model proposed by Barron and Lee [30].

The classical expression for the sound pressure level inside a reverberating room (diffuse field) uses two terms: one for direct sound, which depends on distance to source, and one for reverberant sound, which is independent of the position:

$$L_P = L_W + 10\log\left(\frac{1}{4\pi r^2} + \frac{4}{A}\right) \tag{5}$$

where L_W is the power level emitted by the source; r, the source–receiver distance; and A, the room's absorption, which should be expressed from the measured values of the reverberation time (A=0.161 V/T).

It is appropriate to replace the power level with the acoustic level produced by an omnidirectional source in a free field at a distance of 10 m (L_{P0}= 10 log(W/400π)), so that the levels measured are found at a small interval above the reference level. Therefore, eqn (5) can be written as follows:

$$L_P - L_{P0} = 10\log\left(\frac{100}{r^2} + \frac{31200T}{V}\right) \tag{6}$$

In practice, it is difficult to accept the diffuse field hypothesis and, therefore, uniformity of the reverberant field. Barron and Lee analyse the behaviour of the reverberant field (averaged for mid-range frequencies, the 0.5, 1 and 2 kHz bands) for a group of multiuse rooms whose behaviour cannot be considered diffuse, mainly because most of the absorption is due to the audience surface. To predict the total pressure levels and the indices (such as clarity and definition) that measure the relationship between early and late energy, they propose that the sound energy received at each point be divided into three components, all of which depend on r: direct sound (d); the first reflected sound, with a delay less than 80 ms (e_r); and late reflected sound, with a delay greater than 80 ms (l). With L_{P0}, the relations they propose for the levels are:

$$L_P - L_{P0} = 10\log(d + e_r + l) \tag{7}$$

where:

$$d = \frac{100}{r^2},\quad e_r = \frac{31200T}{V}\mathrm{e}^{-\frac{0.04r}{T}}\left(1 - \mathrm{e}^{-\frac{1.11}{T}}\right),\quad l = \frac{31200T}{V}\mathrm{e}^{-\frac{0.04r}{T}}\mathrm{e}^{-\frac{1.11}{T}} \tag{8}$$

which also implies a sound attenuation according to the distance from the source for the reverberant field (e_r+e_l) of –1dB/10m, approximately.

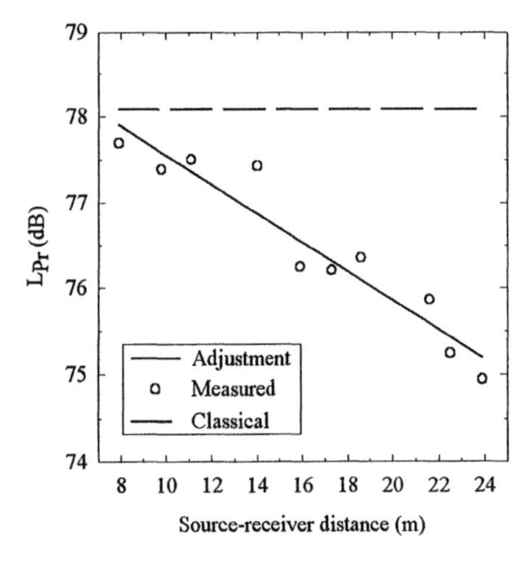

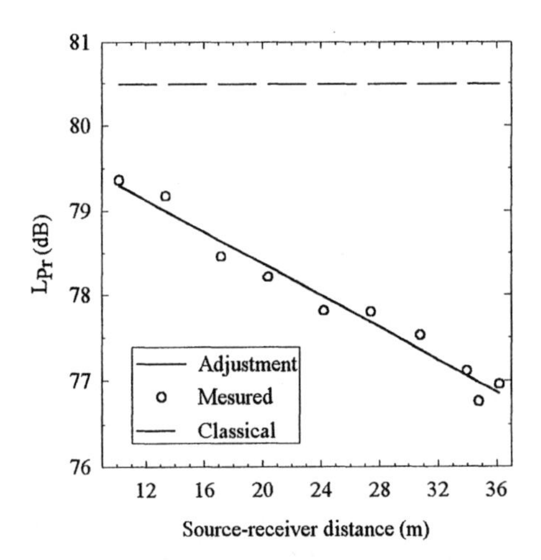

Figure 25. Attenuation of overall L_{Pr} over distance for Santa Catalina Church (left) and for the main nave of Santa Marina (right).

The *in situ* measurements that we carried out suggest that, although the previous model may be appropriate for the overall levels for this type of church (Sendra and Zamarreño [31]), it should be clarified whether this model will be used to predict the behaviour of the frequency in octave bands.

Taking the T values into account and considering that in this type of room there are no particularly absorbent surfaces, we might expect the interior field to match the diffuse model better than the rooms analysed by Barron. However, the form, with its large, smooth surfaces, and its size are not the most appropriate. In fact, Barron's model has already been used to evaluate the impact of an intervention using Schröeder's diffusers in a church (Desarnaulds and Monay [32]).

Our objective then is to analyse how the reverberant field varies over distance, for both the overall levels and for each one of the frequency bands of interest. Therefore, to begin with, we have determined the overall reverberating level, L_{Pr}, at each point based on the measurements carried out, subtracting the energy contributed by the direct field. This is determined based on the (known) transmitter output power and on the source–receiver distance. A linear equation over distance that describes the law of attenuation is adjusted to the values obtained using the least square's procedure. When the correlation of the regression was not suitable, we found that it increased considerably if some of the points situated in the aisles were excluded, mainly the points situated near the entrances of the chapels or side annexes.

Figure 25 shows the data for Santa Catalina and Santa Marina Churches. The regression equations are, respectively:

$$L_{\mathrm{Pradj}} = -0.09r + 80.26 \quad \mathrm{dB} \tag{9}$$

$$L_{\mathrm{Pradj}} = -0.09r + 80.26 \quad \mathrm{dB} \tag{10}$$

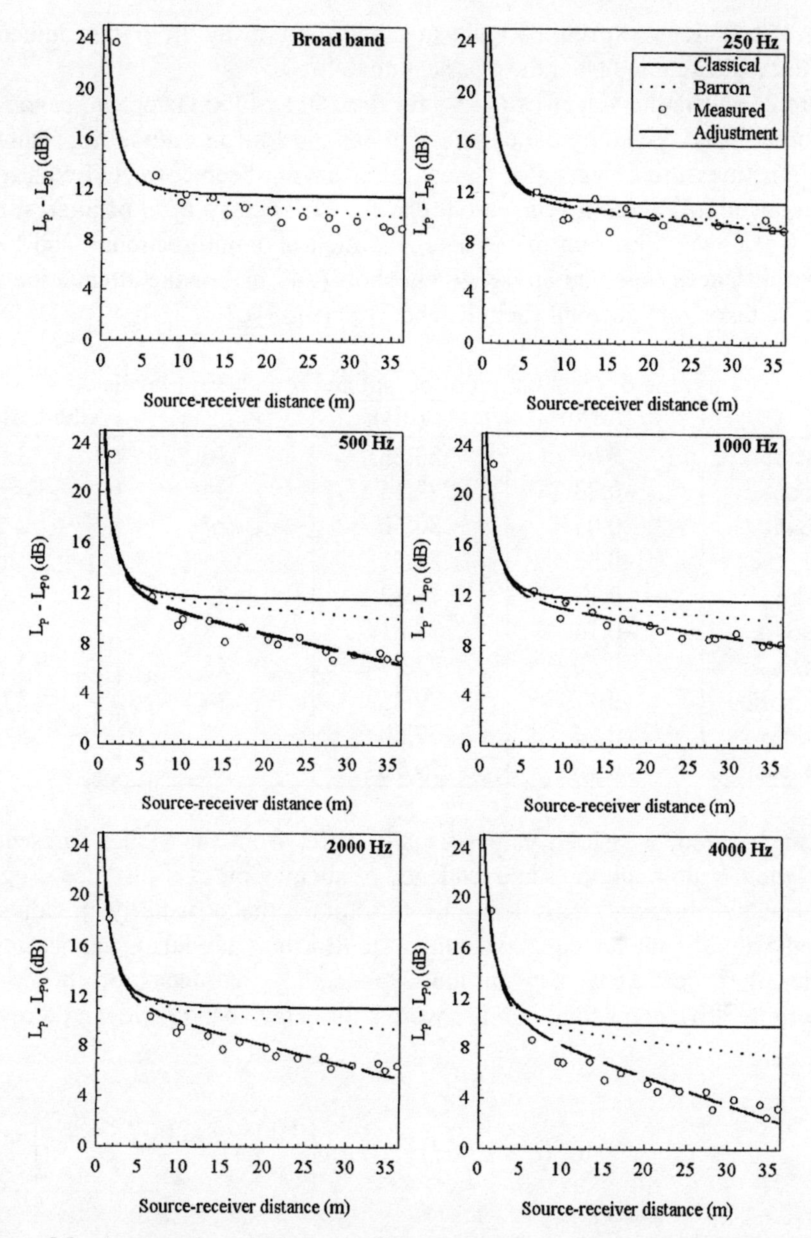

Figure 26. Attenuation of L_P over distance in Santa Marina: comparison of Barron's model and regression curves by octave band.

These equations suggest attenuations for broadband levels of around 1.7 and 0.9 dB/10m, respectively, which give rise to 3 dB in the most distant areas in both churches with respect to the values that would be obtained if we used the classical model. Table 6 summarizes the coefficients of the adjustments made for each church and the attenuations expected in the areas furthest away from the audience with respect to those obtained from the classical model.

Figure 26 shows the values of $L_P – L_{P0}$ for the 250 to 4000 Hz octave bands and also for the broad band, given by the classical model and Barron's model for Santa Marina Church. The measured levels also appear, after having been corrected with respect to reference level L_{P0}. These levels were found for each octave band of interest based on the (known) power spectrum of the source, which is omnidirectional, and bearing in mind that distances involved are relatively short (<40 m) for the attenuation of direct sound to be taken into account (Schultz and Waters [33]).

Table 6. Attenuation of broadband reverberant levels.

Church	m (dB/m)	n (dB)	r_{max} (m)	Attenuation (dB)
Santa Marina	–0.09	80.26	36	3.5
San Vicente	–0.23	77.15	24	4.5
San Julián	–0.05	80.60	27	2.5
San Gil	–0.33	81.58	24	5.0
San Pedro	–0.17	78.82	22	3.0
San Marcos	–0.14	78.81	28	3.1
San Esteban	–0.20	77.97	25	2.8
Santa Catalina	–0.17	79.25	24	3.0
San Isidoro	–0.24	77.50	27	5.0
San Sebastián	–0.16	79.10	23	3.4

As can be seen, measured values usually differ from the values foreseen by the classical model, although the same tendency of attenuation over distance suggested by Barron can be observed. This leads us to explore the possibility of adjusting the measured values with an equation similar to Barron's model, using as adjustment parameter the coefficient of r in the exponential, by means of the Marquard–Levenberg iterative algorithm, in other words, for octave band i, we would rewrite eqn (7):

$$L_P - L_{P0} = 10\log(d + e_r + l) = 10\log\left(\frac{100}{r^2} + \frac{31200T}{V}\mathrm{e}^{-\frac{\beta_i \cdot r}{T_i}}\right) \quad (11)$$

This same procedure was followed for each church. Table 7 summarizes the values of the adjusted factor for each band in each church.

The exponential is adjusted to obtain a suitable expression for e_r and l for subsequent use in evaluating other energy parameters analytically, in addition to the stationary acoustic levels, the main objective of our project. This will allow us to develop a generic model by modifying Barron's model to describe the acoustic field for this type of room. This model could be used in future renovation interventions in which acoustic criteria are taken into account. At present, we are researching the scope of this possibility.

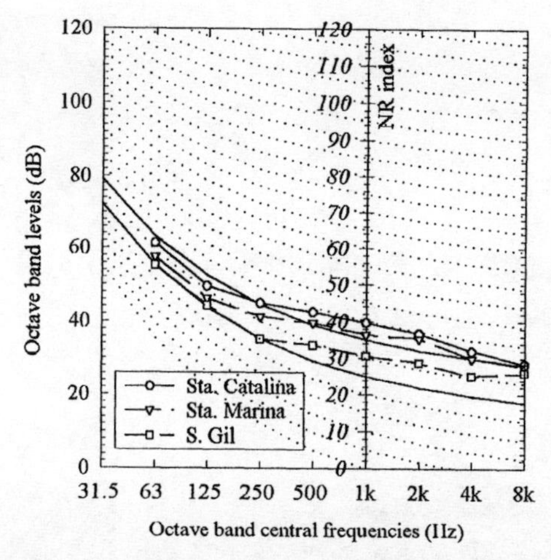

Figure 27. Background noise and NR curves.

Table 7. Adjusted parameters, β_i (s m^{-1}), for a proposed modified Barron's model.

Church	250 Hz	500 Hz	1000 Hz	2000 Hz	4000 Hz
Santa Marina	0.05	0.13	0.09	0.13	0.13
San Vicente	0.06	0.08	0.07	0.10	0.10
San Julián	0.05	0.11	0.06	0.13	0.13
San Gil	0.07	0.13	0.10	0.14	0.13
San Pedro	0.04	0.09	0.07	0.11	0.11
San Marcos	0.10	0.14	0.10	0.15	0.20
San Esteban	0.05	0.10	0.08	0.11	0.12
Santa Catalina	0.05	0.12	0.08	0.12	0.11
San Isidoro	0.06	0.11	0.08	0.12	0.12
San Sebastián	0.04	0.09	0.05	0.09	0.10

3.5 Background noise

We have not considered the systematic evaluation of the acoustic isolation parameters of the most important design elements mainly because the windows and doors are loosely fitted and provide little insulation.

However, we decided to carry out systematic measurements of background noise, which is produced mainly by vehicular traffic, with the procedure described for the sound distribution tests. In fact, the diagram is the same one used in Figure 24, except that background noise takes the place of the source signal. Since this signal is not stationary, the recording and analysis time ranges from 10 to 15 min. In some situations, we took advantage of the capabilities of the data acquisition and analysis system and carried out simultaneous noise measurements at different points.

Figure 28. Church of the Hospital de las Cinco Llagas: (a) aerial view; (b) plan; (c) church interior before renovation; (d) church interior after renovation.

A linear average was performed during the recording time, both for octave bands and for 1/3-octave bands. In order to evaluate the interior noise, the spectra measured were compared with the NR (noise reduction) curves. Figure 27 shows some of the noise spectra measured and highlights the range of the recommended NR curves for this type of room.

These results show average values, which means that the levels may be much higher at certain times, for example when a vehicle goes by, and have a negative effect on the room's acoustic behaviour.

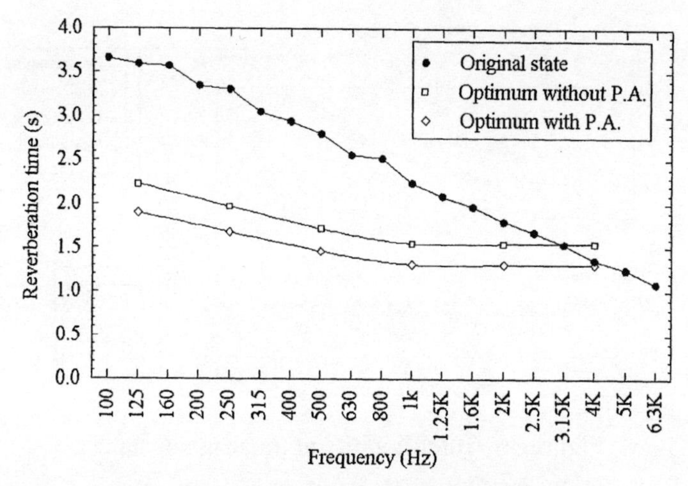

Figure 29. Reverberation at the original state.

4 Three acoustic rehabilitation projects in churches

Now we will examine three of our acoustic rehabilitation projects in churches that have been renovated and given a new use. An analytical methodology similar to that described earlier for Seville Mudejar–Gothic churches is used.

4.1 Conversion of the church of the Hospital de las Cinco Llagas to the general assembly building for the Andalusian Parliament

One of the most important rehabilitation interventions carried out in recent times in Andalusia, although still unfinished, is conversion of the old Hospital de las Cinco Llagas to the seat of the Andalusian Parliament. The building, designed by architect Hernán Ruiz, is a magnificent example of Renaissance architecture (Figure 28a).

The church has a Latin cross floor plan with small side chapels (Figure 28b) covered with stone vaults supported by walls and pilasters of the same material (Figure 28c). It has an approximate volume of 12,000 m^3, an average height of 20 m, an interior surface of 6000 m^2 with 450 m^2 of useful floor space, and approximate capacity of 300 people. These figures give the following proportions: 40 m^3 per person (rather than the standard four or five cubic metres); a volume/area ratio of 26 m^3/m^2 (rather than the suggested 8 cubic metres); and a low, but not disproportionate, occupation density of 1.5 m^2/person.

The rehabilitation project was drawn up by architects A. Jiménez and P. Rodríguez. Our group worked closely with them from the preliminary planning stage, allowing us to combine design and functional decisions with the requirements of acoustic conditioning (Figure 28d).

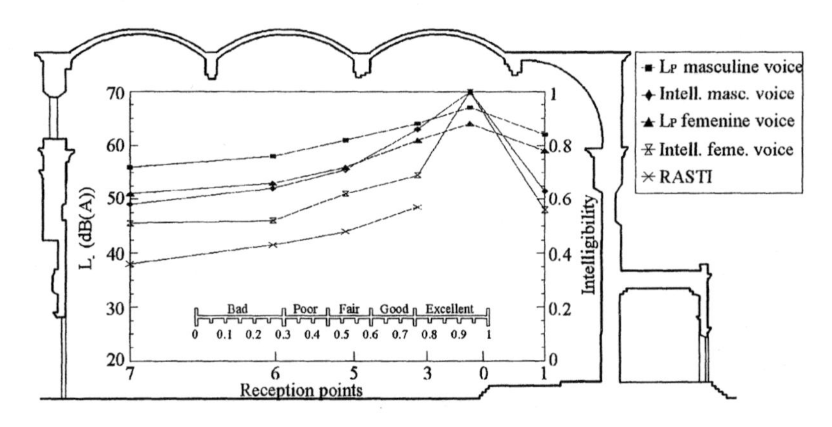

Figure 30. Intelligibility at the original state.

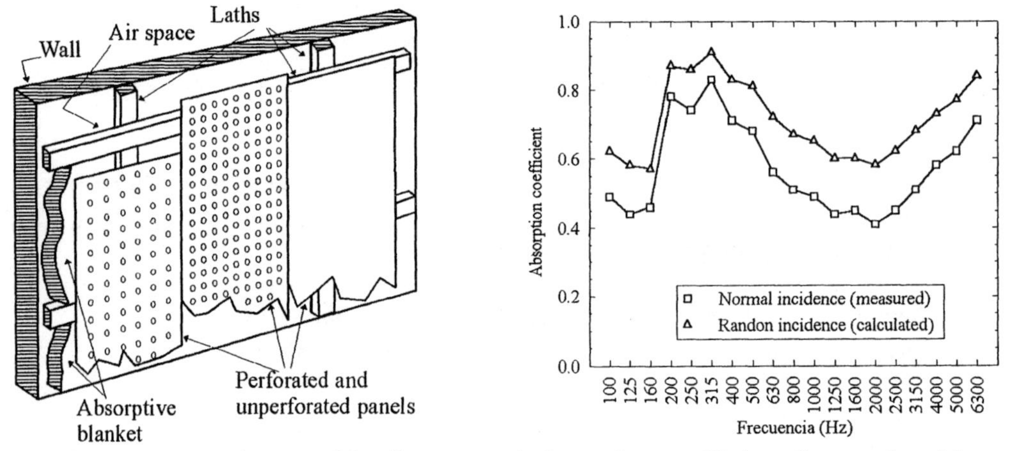

Figure 31. Panels assembly diagram and absorption coefficient for panels with 5 mm perforations.

4.1.1 Original state

The original acoustic conditions of the buildings were exhaustively analysed based on complete *in situ* measurements taken in the empty church. We are including a summary of the most significant data: reverberation times (Figure 29), and subjective intelligibility and RASTI index tests performed along the axis of the nave (Figure 30). As can be observed, the reverberation times were excessive and intelligibility could be described as “poor”. The analysis of the modulation transfer functions revealed that reverberation was the main cause of the deterioration of the RASTI index values.

Levels between 32 and 34 dB(A) were obtained for background noise inside the church, while on the nearest street, which had very dense traffic, levels of between 74 and 82 dB(A) were measured. Its heavy walls and the fact that the church is built within one of the hospital's courtyards helped keep these values at acceptable levels.

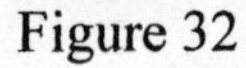
Figure 32

Figure 33

4.1.2 Main decisions adopted for acoustic rehabilitation

Since the reverberation time values were very high and the volume of the room, for evident architectural reasons, could not be reduced, a decision was made to increase acoustic absorption, principally at medium and low frequencies, using absorbent membranes and perforated panels. In this fashion, perforated and solid panels backed with a 5 cm layer of rock wool were combined to situate the resonance frequencies in the 125–400 Hz range. These panels were held in place by wooden laths, leaving a 25 cm air-space (Figure 31). These absorbent elements, with a final layer of velvet, were placed in all the side chapels (to take the place of the original retables and to help conceal the air-conditioning system) and, to an even greater extent, along sidewalls of the triforium or upper gallery (Figure 32). The main altarpiece was covered with a very heavy tapestry, and a heavy carpet was laid on the marble floor where no wooden furniture was to be placed. These elements were meant to absorb and diffuse the sound (Figure 33). Altogether, 600 square metres of absorbent panels and 400 square metres of carpet were installed.

4.1.3 Final state

Figures 34 and 35 show the results of the final acoustic measurements (after rehabilitation): reverberation time and RASTI index (the RASTI index measurements were taken with the PA system on and off). The original values are also shown in these figures for comparison. The manifest improvement of the acoustical conditions makes the room very appropriate for its new use.

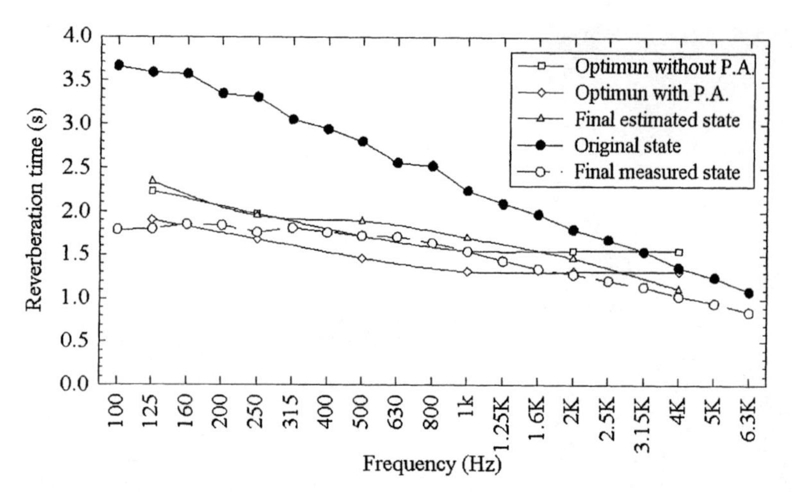

Figure 34. Reverberation from original to final state.

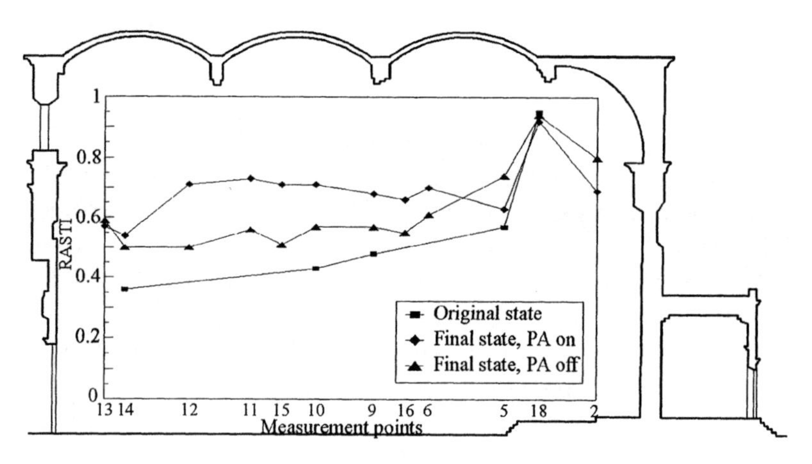

Figure 35. RASTI index from original to final state.

4.2 Rehabilitation of the San Francisco de Baeza Monastery Church as a centre for culture and music

This outstanding Renaissance church, which is attributed to architect Andrés de Vandelvira, was in ruins after the Spanish Civil War and stayed that way until the 1980s, when it was rehabilitated as a cultural centre, mainly for lectures and concerts. This rehabilitation was carried out without taking into account acoustic criteria. Very high reverberation times, which even produced echoes, inappropriate sound distribution, and a great deal of background noise due to the use of curtain walls with insufficient acoustic insulation soon became evident.

The church has a cross-shaped floor plan with three wings of similar size in the middle of which rises a platform that serves as a stage (Figure 36a). It has a vaulted

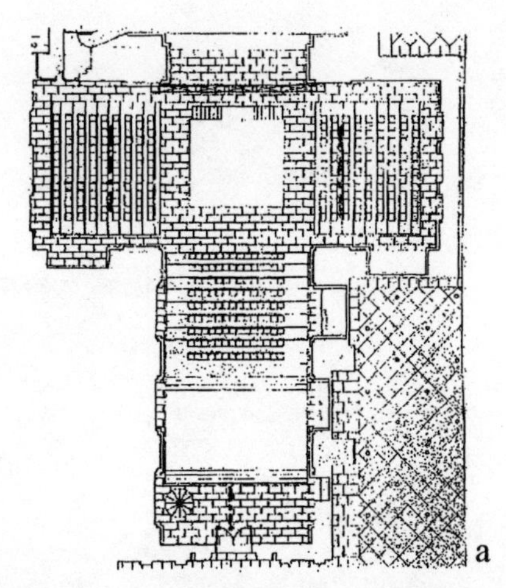

Figure 36. San Francisco de Baeza Church: (a) plan; (b) inside before renovation.

stone roof supported by walls and pilasters made of the same material (Figure 36b). It has a volume of approximately 7555 m^3, an average height of 17 m, and 2025 m^2 of interior surface, of which 396 m^2 is useful space, and is designed to seat 350 people. These figures give the following proportions: 21 m^3 per person (rather than the standard five or six cubic metres); a volume/area ratio of 26 m^3/m^2 (rather than the suggested 8 m^3); and a low occupation density of 0.8 m^2/person.

At the request of the regional government, an acoustic reconditioning plan, or perhaps we should call it an acoustic rehabilitation plan, was drawn up. This plan takes our advice into account. This case is representative of the situation we pointed out in the introduction: once the rehabilitation is finished, a new project has to be undertaken to correct the acoustic problems not taken into account in the original design.

4.2.1 Original state

Therefore, acoustic measurements of the finished building were performed, although from the point of view of our project the finished building was the original state of an unfinished project. Figure 37 shows the reverberation times. As can be seen, the recommended values are greatly exceeded. The intelligibility measured was also deficient, with RASTI index values of between 0.25 and 0.30 where the audience would be seated.

4.2.2 Principal decisions adopted for acoustic rehabilitation

In addition to the negative acoustic conditions at the start, it was also difficult to design corrective measures that would not require the total or partial destruction of the newly finished renovation work. A decision was made to work together with the architects who carried out the renovation, and the following measures were proposed:

- Substitution of heavily padded seats for existing stone and wood tiers (Figure 38).

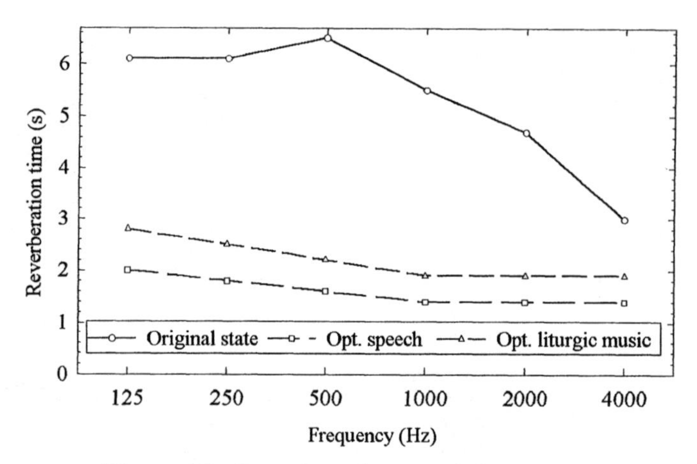

Figure 37. Reverberation at original state.

- Installation of motorized extensible awnings that can temporarily cover the ceiling of the church at the level of the cornice. This measure, besides correcting the reverberation time at certain frequencies, has a positive influence on sound distribution because it reduces the harmful effect of sound reflected from the vaults (Figure 39).
- Separation of the choir from the rest of the auditorium by means of a heavy, mobile curtain. This measure, together with the previous one, forms a smaller room within the original volume, thus helping to eliminate the long reflections and echoes and improving intelligibility (Figure 40).
- Installation of a sheet of absorbent material about 80 cm in front of the glass curtain wall at the crossing, behind what would be the stage (Figure 41).

4.2.3 Final state

A graph is included (Figure 42) comparing the original reverberation times and the reverberation times measured after the intervention. The nature of the correction produced can be observed. RASTI index values were much improved, ranging between 0.45 and 0.55 for all parts of the audience, but could not called outstanding. A final NR index reading of 44 was obtained for background noise, much higher than desired because the supplementary acoustic insulation we recommended was not installed due to economy.

Although the original conditions were difficult because of the church's acoustic conditions and because of the limited range of corrective measures that could be used in a newly renovated building, the results obtained prove that the solutions were suitable. However, other suggested measures that were not carried out would have improved the results. In any case, the building can now be considered an appropriate venue for cultural activities.

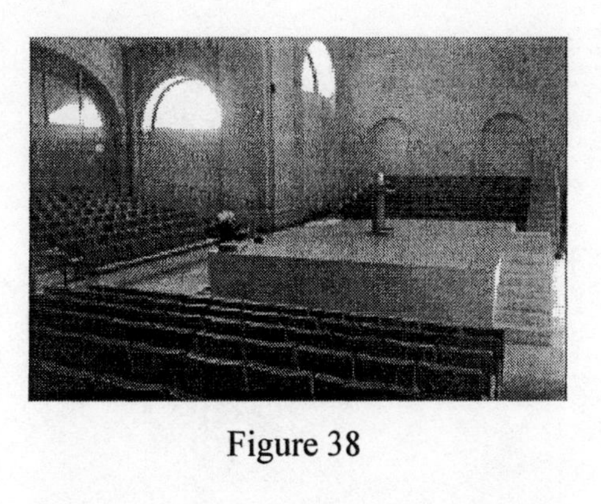

Figure 38

Figure 39

Figure 40

Figure 41

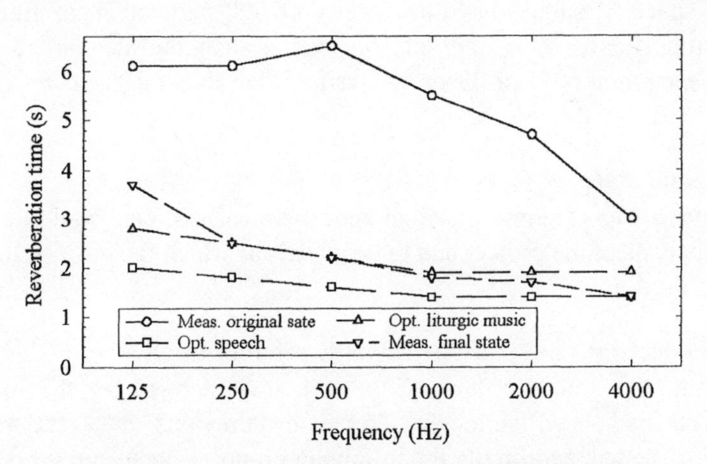

Figure 42. Reverberation in original and final states.

4.3 Rehabilitation of the Carmen Church in Vélez-Málaga for use as a theatre

The present-day Carmen Theatre in Velez-Málaga is the result of a rehabilitation and adaptation plan drawn up by architect A. González, who took our advice on acoustics into account. This plan was realized by the Autonomous Government of Andalusia. The hall is now used mainly for theatre, although it may also be used for concerts

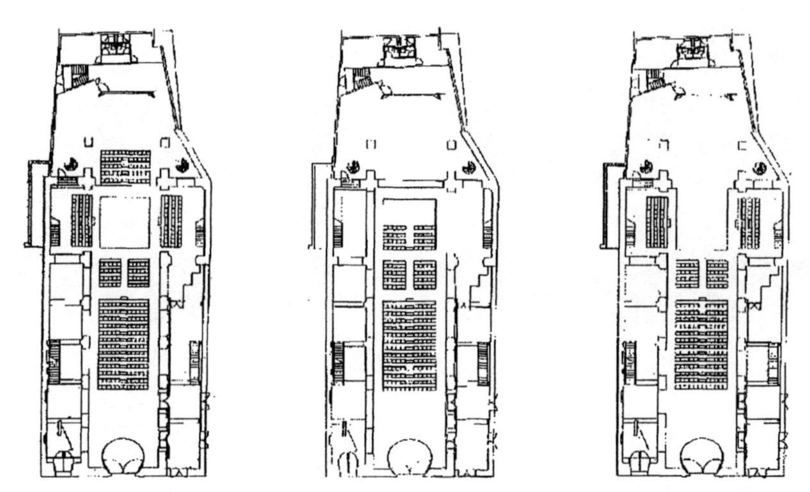

Figure 43. Different room layouts: modern theatre; classical theatre; musical with an extended stage.

(Figure 43). This Baroque church was part of the Carmen Convent, although it was the object of unfortunate alterations in the course of the twentieth century to adapt it for other uses, especially for cinema.

This three-nave church has a vaulted roof, although the room that was rehabilitated only occupies the central nave (Figure 44). This room has an approximate volume of 4760 m^3, an average height of 15 m, and an interior surface of 2545 m^2 with 380 m^2 of useful floor space. It should have a capacity of 300 people. These figures give the following proportions: 16 m^3 per person (rather than the standard 5 or 6 m^3); a volume/audience area ratio of 29 m^3/m^2 (rather than the suggested 8 m^3); and a low occupation density of 0.6 m^2/person.

4.3.1 Original state

Since the church was in ruins, no initial acoustic measurements could be taken, and a theoretical analysis of the project had to be made, for which the RAYNOISE computer program was used.

4.3.2 Main decisions adopted for acoustic rehabilitation

From the start, the architect decided to work toward restoring the interior of the church, which had been seriously altered by previous interventions. Different operations were tested, and finally the following proposals were carried out:

- A room was defined within the church and a large surface of heavy, gathered, mobile velvet curtains was hung to mitigate the reverberation times (Figure 45).
- The room has a double glass curtain wall that favours acoustic insulation. The outside wall is made of 6+12+3+3 mm laminated glass, and the inside partition is made of 10 mm tempered glass. The side wall may be covered by the curtain (Figure 46).

Figure 44

Figure 45

Figure 46

Figure 47

- The nave is covered with a vault, but this vault is made of two sheets of laminated wood separated by laths and filled with mineral wool to provide thermal and acoustic insulation (Figure 47).

4.3.3 Final state

Once the rehabilitation work was finished, acoustic measurements were taken *in situ* with the curtains deployed in different positions. Figure 48 shows the results of the reverberation times. The values found are very close to the recommended values for these uses.

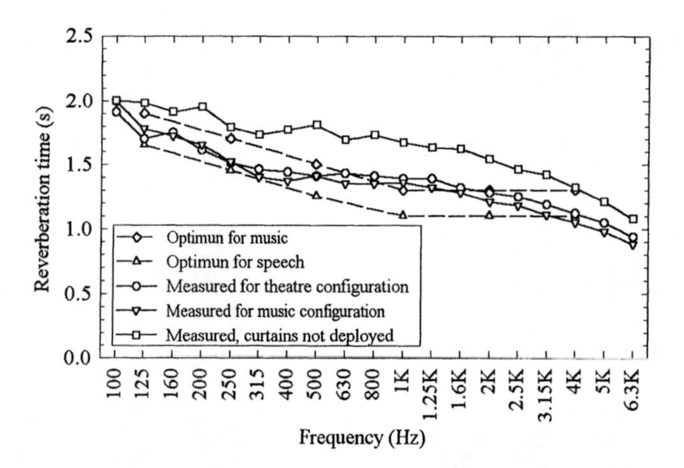

Figure 48. Reverberation times for different uses.

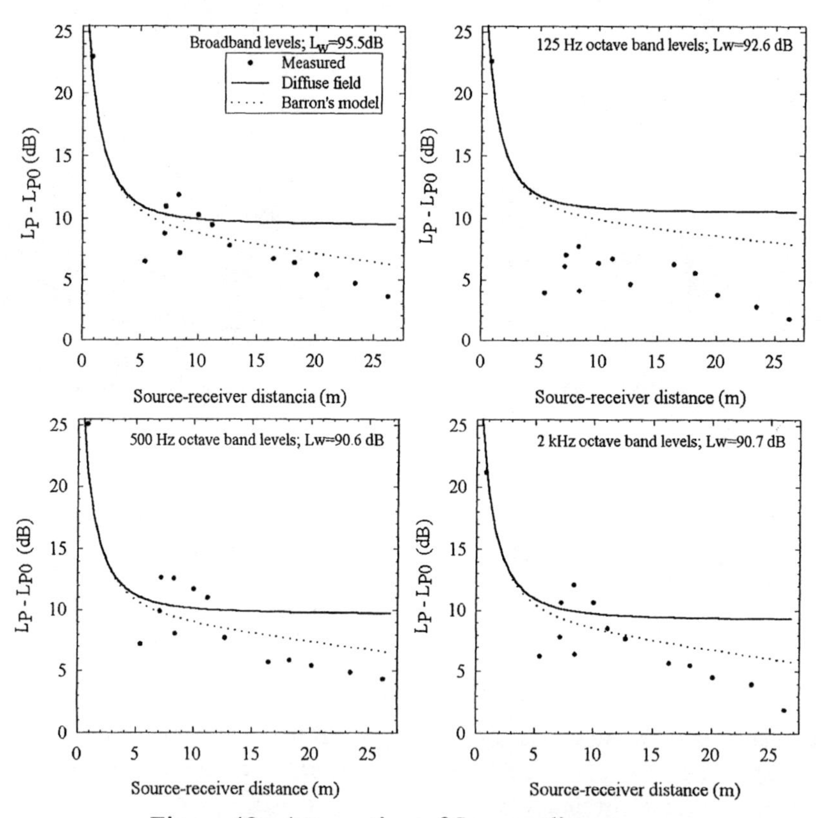

Figure 49. Attenuation of L_P over distance.

The values found for the RASTI index ranged between 0.45 and 0.60 at different points in the room, which allows us to classify intelligibility as "fair". The background noise was measured with the HVAC system on and off. With the system running, an

NR index of less than 30 was obtained, which is a very appropriate value for the intended uses.

Finally, Figure 49 shows the distribution of sound levels determined by broadband and octave band measurements and compares them with the levels obtained with the classical diffuse sound field model and Barron's model for the reverberant field, commented on in detail in the study on Mudejar churches in Seville.

Therefore, we can assert that, in spite of the original difficulties, we were able to harmonize the architect's stylistic criteria and formal requirements with the functional acoustic objectives. In general, we can classify the room's acoustic behaviour as suitable for the intended uses.

References

[1] Navarro, J., and Sendra, J.J., La iglesia como lugar de la música, *Actas del Primer Congreso Nacional de Historia de la Construcción*, Madrid, pp.381-387, 1996.

[2] Sendra, J.J., and Navarro, J., El concilio de Trento y las condiciones acústicas en las iglesias, *Actas del Primer Congreso Nacional de Historia de la Construcción*, Madrid, pp. 485-490, 1996.

[3] Sendra, J.J., and Navarro, J., *La evolución de las condiciones acústicas en las iglesias: del Paleocristiano al Tardobarroco*, Instituto Universitario de Ciencias de la Construcción, Universidad de Sevilla, 1997.

[4] Abraham, G., *The Concise Oxford History of Music*, Oxford University Press, 1979.

[5] Sendra, J.J., Zamarreño, T., and Navarro, J., An analytical model for evaluating the sound field in Gothic Mudejar churches, *Proc. of the Second Int. Conf. on Computational Acoustics and its Enviromental Applications*, eds. C.A. Brebbia, J. Kenny and R.D. Ciskowsky, Computational Mechanics Publications, Southampton, pp. 139-148, 1997.

[6] Sendra, J., Zamarreño, T., and Navarro, J., Acoustical behavior of churches: Mudejar Gothic churches, *Proc. of the 16th Int. Congress on Acoustics and 135th Meeting Acoustical Society of America*, Seattle, 1998.

[7] Shankland, R.S., and Shankland, H.K., Acoustics of St. Peter's and Patriarcal Basilicas in Rome, *J. Acoust. Soc. Am.*, **50**, pp. 389-396, 1971.

[8] Riguetti, M., *Historia de la liturgia latina*, Biblioteca de Autores Cristianos, Madrid, p. 231, 1956.

[9] Jungmann, J.A., *El sacrificio de la Misa. Tratado histórico-litúrgico*, Biblioteca de Autores Cristianos, Madrid, p. 171, 1951.

[10] Jungmann, J.A., op. cit., p. 582.

[11] Sendra, J., and Zamarreño, T., El campo sonoro en las iglesias gótico-mudéjares con cubierta de madera: aplicación del modelo de Barron, *Actas de las Jornadas Nacionales de Acústica: Tecni-Acústica'95*, A Coruña, pp. 87-90,1995.

[12] Benedetti, S., *Fuori dal clasicismo*, Bonsignori Ed., p. 87, 1993.

[13] Benedetti, S., op. cit., pp. 17-20.

[14] Sendra, J.J., Navarro, J., and Zamarreño, T., La forma de cubrir las iglesias y su relación con las condiciones acústicas, *Proc. del Primer Congreso Iberoamericano de Acústica*, Florianópolis, pp. 495-498, 1998.

[15] Wittkober, R., *Architectural Principles in the Age of Humanisme*, Academy Editions, London, Appendix I, 1952.

[16] Forsyth, M., *Buildings for Music*, Cambridge University Press, London, p. 3, 1985.

[17] Rodríguez, A., *Bartolomé de Bustamante y los Orígenes de la Arquitectura Jesuítica en España*, Institutum Historicum S.I, Roma, docum. 39, p. 367. ARSI, Hisp. 111, f. 78r, 1967.

[18] Pirri, P., *Giovanni Tristano e Primordi della Architettura Gesuitica*, Institutum Historicum S.I., Roma, pp. 146, 228-229, 1995.

[19] Pirri, P., op. cit., p. 147.

[20] Vallery-Radot, E., and Lamalle, E., *Le recueil des plans d'édifices de la Compagnie de Jésus conservé à Bibliothèque Nationale de Paris,* Institutum Historicum S.J., Roma, 1960.

[21] Schnoebelen, A., Performances Practiques at San Petronio in the Baroque, *Acta Musicológica*, **P. XLI**, pp. 37-55, 1969.

[22] Angulo, D., *Arquitectura mudéjar sevillana de los siglos XIII, XIV y XV,* Ayuntamiento de Sevilla, p. 32, 1983.

[23] Borrás, G.M., and. Sureda, J., *Historia del arte español (tomo III),* Planeta, Barcelona, 1995.

[24] Zamarreño, T., and Algaba, J., Teoría y práctica de la medida de los tiempos de reverberación cortos, *Rev. Acúst.*, **XVIII,** (3, 4), pp. 71-73, 1997.

[25] Schröeder, M.R., New method of measuring reverberation time, *J. Acoust. Soc. Am.*, **37**, pp. 409-412, 1965.

[26] Sendra, J.J., Zamarreño, T., Navarro, J., and Algaba, J., *El problema de las condiciones acústicas en las iglesias: principios y propuestas para la rehabilitación.* Instituto Universitario de Ciencias de la Construcción, Universidad de Sevilla, 1997.

[27] Knudsen, V.O., and Harris, C.M., *Le Projet Acoustique en Arquitecture*, Dunod, Paris, 1957.

[28] Sendra, J.J., Zamarreño, T., Navarro, J., and Algaba, J., op. cit., cap. II.

[29] Hugast, T., and Steeneken, H.J.M., A review of the MTF concept in room acoustics and its use for estimating speech intelligibility in auditoria, *J. Acoust. Soc. Am.*, **77**, (3), pp. 1069-1077, 1985.

[30] Barron, M., and Lee, L.J., Energy relations in concert auditoriums I, *J. Acoust. Soc. Am.*, **84**, pp. 618-628, 1988.

[31] Sendra, J., and Zamarreño, T., El campo sonoro en las iglesias gótico-mudéjares con cubierta de madera: aplicación del modelo de Barron, *Proc. of the Jornadas Nacionales de Acústica: Tecni-Acústica'95*, A Coruña, pp. 87-90, 1995.

[32] Desarnaulds, V., and Monay G., Utilisation des diffuseurs de Schröeder dans les salles polyvalentes à plafond haut, *Proc. of the 11th Int. FASE Symposium*. Valencia, pp. 15-17, 1994.

[33] Schultz, T.J., and Waters, B.G., Propagation of sound across audience seating, *J. Acoust. Soc. Am.*, **36**, pp. 885-896, 1964.

WIT PRESS

Structural Studies, Repairs and Maintenance of Historical Buildings

Editors: **S. SÁNCHEZ-BEITIA**, University of the Basque Country, Spain and **C.A. BREBBIA**, Wessex Institute of Technology, UK.

This book comprises the papers presented at the Fifth International Conference on Structural Studies, Repairs and Maintenance of Historical Buildings, held in 1997. The following wide variety of topics are covered: History and Architecture; Monitoring and Testing; Computer Simulation; Deterioration and Protection of Materials; Material Evaluation and Restoration; Retrofitting; Different Types of Structures; Domes; Masonry; Seismic Behaviour and Vibrations; Repairs and Strengthening; Case Studies; and Heritage as a Factor of Development.

ISBN: 1-85312-466-4 **1997**
696pp **£198.00/US$299.00**

Towers and Domes

F. ESCRIG, University of Sevilla, Spain.

This highly illustrated book is designed to appeal to those who are interested in the history of architecture as well as those actively involved in the field. Detailing the evolution of towers and domes from a structural viewpoint, it is not written as a detailed study on technical or mathematical concepts, but rather as two essays running parallel, one textual, the other graphic.

Contents: The Architecture for Survival; The Great Architecture of Nature; Piled Stones; Arguments of Authority; Inventing Architecture Again; Petrified Woods; Against Vegetal Architecture; Dreams of Reason; Domesticated Nature.

ISBN: 1-85312-437-0 **1998**
120pp **£59.00/US$95.00**

Historical Buildings of Iran

Their Architecture and Structure

M.M. HEJAZI, Queen Mary and Westfield College, University of London, UK.

The first authoritative work to investigate the historical buildings of Iran from the perspective of structural engineering, this volume includes a chronological description of architectural styles as well as chapters on traditional construction materials, arches, vaults, domes, minarets and water works. The author also proposes several historical structures as research topics for future work.

Contents: Preface; Introduction; Ars sine scientia nihil: Knowledge in Persian Architecture; Construction Materials of Historical Structures; Architectural and Structural Features of Persian Historical Buildings; Arches; Vaults; Domes; Minarets; Water-works; Maintenance Comments on the Wooden Structure of the Ali Qapu Building; Proposal for Future Research into Historical Structures; References; Index.

ISBN: 1-85312-484-2 **1997**
168pp **£67.00/US$99.00**